Mobile Video Telephony

Mobile Video Telephony

For 3G Wireless Networks

David J. Myers

McGraw-Hill

New York   Chicago   San Francisco   Lisbon   London   Madrid
Mexico City   Milan   New Delhi   San Juan   Seoul
Singapore   Sydney   Toronto

The McGraw·Hill Companies

Library of Congress Cataloging-in-Publication Data

Myers, David J.
    Mobile video telephony : for 3G wireless networks / David J. Myers.
       p.   cm.
    Includes bibliographical references and index.
    ISBN 0-07-144568-4
    1. Video telephone. 2. Mobile communication systems. 3. Global system
for mobile communications. I. Title.
    TK6505.M94   2004
    621.3845′6—dc22   2004055977

1 2 3 4 5 6 7 8 9 0 DOC/DOC 0 1 0 9 8 7 6 5 4

ISBN 0-07-144568-4

*The sponsoring editor for this book was Stephen S. Chapman and the production
supervisor was Sherri Souffrance. It was set in Century Schoolbook by
International Typesetting and Composition. The art director for the cover
was Handel Low.*

Printed and bound by RR Donnelley.

McGraw-Hill books are available at special quantity discounts to use as pre-
miums and sales promotions, or for use in corporate training programs. For more
information, please write to the Director of Special Sales, McGraw-Hill
Professional, Two Penn Plaza, New York, NY 10121-2298. Or contact your local
bookstore.

For Amanda, Rupert, Helen, and Elizabeth

Contents

Foreword

Mobile Video Telephony is becoming a standard feature of *third-generation* (3G) networks. Prior industry experience with video-conferencing over *integrated services digital networks* (ISDN) and the relative maturity of the service surround for circuit-switched networks compared to emerging packet networks means that circuit-switched 3G-324M video telephony has been adopted for the first releases of 3G networks. The video telephony services that are available today from companies such as 3, NTT DoCoMo and Vodafone, and that other mobile operators are launching, are all based on this standard.

Not only is video telephony a new experience for users, it is also a new paradigm for the mobile operators and engineers that are launching conversational video services. They need to widen their expertise in providing speech telephony to cover the additional layers of protocols and technologies that are needed to extend services to add a video component to the communication experience. The same can also be said for most of the companies that manufacture and supply equipment and handsets to the mobile operators. The success of video telephony services depends on providing a positive user experience. The engineering of user-friendly 3G-324M terminals and services requires understanding of the underlying 3G-324M protocol and of how such systems can be designed and optimized for reliability, interoperability, and high quality.

It has become clear from the training courses that Dilithium Networks runs for its customers that there is a real need for information on 3G-324M, which is not readily available from any single or even a small number of sources. 3G-324M is a complex standard that makes reference to several other standards, most of which run to hundreds of pages themselves, and are written for thoroughness rather than readability.

As long ago as 1993, when David Myers headed the Image Processing Research Group at British Telecom's R&D laboratories, he placed a contract with me (then at Sydney University) to work on low bit-rate video communication. This led directly to my participation in the ITU-T working group that developed the H.324 recommendation. As a result of this

Macchina, the predecessor of Dilithium Networks, was the first company in the world to offer commercial implementations of H.324 and 3G-324M—making them available from 1998 onward.

David Myers' book provides an informed synthesis of the components that make up 3G-324M systems, in sufficient depth to allow those who are working on equipment or services that use the standard to understand how it operates. By providing the necessary background on multimedia communication, he enables people who have not previously worked in this area to quickly come to grips with the details of the standard.

Throughout the book he provides insights from his own experience on why the standard operates as it does. He illustrates approaches to diagnosing problems when they occur. He describes real-world considerations in the design and development of equipment that implements the protocols to produce well-engineered systems with optimized performance.

The book provides easy-to-read explanations of the key building blocks of 3G-324M systems and outlines how mobile video telephony will evolve in the long term, looking at the problems of interworking between different protocols and technologies.

I recommend it as a significant contribution to making the underlying principles of 3G video telephony accessible and understandable to practitioners in industry. It should also serve as an excellent resource for advanced undergraduate and postgraduate courses and research work designed to prepare people for careers in mobile telecommunications.

DR. MARWAN JABRI
Chief Technical Officer, Dilithium Networks;
Member of the Board of Directors
of the International Multimedia
Telecommunications Consortium

Acknowledgments

This book was an opportunity to consolidate the knowledge and experience gained from almost 3 years that I spent helping to take Dilithium Networks from inception to a position of leadership in *third-generation* (3G) video telephony technologies and products.

By inviting me to join him in founding Dilithium Networks, Marwan Jabri gave me the chance to recruit an excellent engineering team. Their skill, commitment, and support made them a great group of people to work with.

I would like to acknowledge the help of Stephen Brown, Raymond Wang, Robert Jongbloed, Sam Georgy, and Albert Wong in providing advice and clarification on different technical questions, and to thank Malcolm Macnaughtan and Raj Pandey for reviewing the complete manuscript and suggesting improvements.

The final manuscript owes much to the work of Steve Chapman, Neha Rathor, and the rest of the team at McGraw-Hill.

Introduction

The idea of video telephony has been with us for many years. Science fiction movies and television shows have consistently depicted it since the 1950s and 1960s, technology gurus have frequently forecast its imminent rise to prominence, yet it has never yet properly made the transition from the development labs to mass-market product: it is a technology that has a long history of predictions of a glorious future behind it. What looks set to finally change this is the deployment of third generation (3G) mobile networks. Mobile operators are pinning their hopes on providing conversational video telephony services to encourage customers to move from their existing second generation (2G) mobile service to the new 3G mobile networks.

Handsets that support video telephony are being produced by a number of major manufacturers. In Europe 3G networks are becoming commercially available and operators are offering mobile video telephony handsets to their customers. In Japan DoCoMo's FOMA-based video telephony service is growing rapidly.

The promise of 3G mobile services is that they will be a significant step toward universal access, providing mobile users with access to all the content and services they are used to getting on the Internet through dial-up, *asymmetric digital subscriber loop* (ADSL) or their office *local area network* (LAN), as well as all the mobile telephony services that they are used to. As customers become familiar with and habituated to using mobile video telephony, they will expect to use similar services over different networks, including their Internet connection (video and *voice over Internet protocol,* or VoIP) and their home telephone. Mobile video telephony may well provide the catalyst for the widespread adoption of video telephony services.

The purpose of this book is to provide a good system-level understanding of mobile video telephony. This is intended to be useful to people working on the deployment of 3G video services, people developing handsets and network equipment for such services as well as a source of information for managers, technical marketing people and people in advanced university courses in this subject, producing new recruits for the industry.

With the intention of putting the rest of the book into context, Chapter 1, "Migrating to Third Generation Mobile Networks," provides an overview of 2G and 3G mobile networks and the proposed approaches to migrating from 2G to 3G networks. The standards used in 3G networks are based on the work of groups such as 3GPP and 3GPP2. They have proposed a set of recommendations that allow carriers to build on their existing investment in 2G infrastructure, to keep costs manageable. They envision that the 3G network itself will follow an evolutionary path, starting with an architecture often referred to as Release 99 (sometimes also known as Release 3) and evolving to Release 5 and later releases. In Release 99 architectures, conversational services that require guaranteed delivery with real-time constraints and no repudiation of billing are provided over the more traditional circuit-switched network. Later releases propose packet-based delivery for such services, but will need *quality-of-service* (QoS) guarantees and reliable billing mechanisms that are at least comparable with those provided for existing circuit switched services. This chapter provides readers who are not familiar with mobile telephony systems with the necessary background to help them understand how video telephony maps onto this evolving mobile network architecture.

Chapter 2, "Basics of Multimedia Communication," is aimed at readers who may not be familiar with multimedia systems. It provides an introduction to the general principles of conversational services that use multiple media types on both switched and packet networks. It discusses the process of establishing and controlling a multimedia call. It also describes how terminals can differentiate between the different media types they receive to decode the data appropriately for presentation to the end user.

Although this book is not intended to cover video and speech coding in any depth, Chapter 3, "Media Coding," introduces the principles behind the digital encoding of video and speech signals. The G.723.1 and *global system for mobile communications adaptive multirate* (GSM-AMR) standards for speech coding and the H.263 and *moving picture experts group-4* (MPEG4) standards for video coding are described in sufficient detail to enable the reader to understand the systems aspects of handling these media types that are presented in later chapters.

Video telephony services on Release 99 3G networks are based on 3G-324M, a derivative of the H.324 recommendation for low-bit-rate media communication of the International Telecommunications Union (ITU). H.324 was originally intended for video telephony over the *general switched telephone network* (GSTN). Through a series of annexes it has been extended to make it suitable for use on mobile devices and networks. When implemented with these annexes it is often referred to as H.324M. Chapter 4, "Video Telephony over Mobile Switched Networks," describes this standard and the closely related 3G-324M recommendation that has been adopted by 3GPP.

The 3G-324M and H.324M recommendations in their turn reference other recommendations. These cover aspects such as video and speech coding, multiplexing and demultiplexing the media (combining media into a single digital bit stream and extracting it when it is received) and session establishment. The ITU-T H.223 Multiplexing protocol is described in Chapter 5, "The H.223 Multiplexer in Detail." Speech-based mobile telephony requires that call signalling be used to establish an end-to-end link or bearer. Once this is established, sending digitally encoded speech over this bearer is fairly straightforward. In contrast, for multimedia communication to take place a series of negotiation steps are needed after the bearer has been established to complete the process of session establishment. The control protocol used in 3G-324M and H.324M is described in depth in Chapter 6, "H.245 Command and Control in Detail."

Chapters 4 to 6 are intended to provide a complete description of the underlying principles of 3G-324M. Chapter 7, "Session Walkthrough," consolidates and builds on what has been presented in these chapters by analyzing an idealized call between two mobile video telephony handsets.

Many factors will ultimately decide whether circuit-switched mobile video telephony is a success. 3G-324M is considerably more complex and sophisticated than standard voice telephony. The implementation of 3G-324M on handsets, with limited computational power and battery life, is a significant challenge. Conformance of the implementation to the standards, to ensure that handsets from different manufacturers will operate with each other and with network equipment is also a significant task. Diagnosing problems that may arise with 3G-324M equipment and services requires special tools. These areas are discussed in Chapter 8, "Implementation Issues."

As 3G networks evolve, conversational services will also be carried over packet networks. Because of this, the technology used to provide video telephony is expected to change significantly. For Release 5 networks, it is proposed that handsets will be based on the Internet Engineering Task Force (IETF) Session Initiation and Session Description Protocols (SIP/SDP). Work in this area is less advanced than for 3G-324M, where handset manufacturers have already launched 3G-324M products. Chapter 9, "Video Telephony over Mobile Packet Networks," describes the concepts behind the future development of mobile video telephony based on SIP.

In the early stages of 3G deployment there will be islands of 3G within a largely 2G and 2.5G global network; the operators will have to handle this by building video-awareness into the network, allowing video-enabled and voice-only terminals to interoperate. Therefore, enabling 3G video telephony services is as much about building 3G-324M awareness into the network as it is about providing 3G-324M handsets.

For mobile video telephony to develop as a mass-market product a critical mass of users must be quickly established; once you have bought your mobile videophone you need someone to call. To facilitate this, mobile video telephony services must provide access to as wide a variety of video-enabled end users as possible. This can be assisted by enabling 3G video telephony customers to call people on Internet-based H.323 and SIP video and VoIP terminals and to call *integrated services digital network* (ISDN) based H.320 video telephone users. Providing this interoperability will make the emerging mobile video telephony service more attractive, particularly to corporate customers.

Customers of video telephony services will expect to be able to access a similar range of services to those they get with their voice service. These include roaming, call forwarding, multipoint calling, and mailbox services. The existing infrastructure that provides these services needs to be enabled or enhanced to take account of the video aspect of the call.

Providing interworking and access to supplementary services to achieve the goal of mass-market mobile video telephony requires sophisticated gateways at the edge of the mobile network. These gateways convert control signaling and media coding formats between 3G-324M and a range of other popular protocols including H.323, SIP, and H.320. As Release 5 services become available, such gateways will also allow legacy circuit-switched Release 99 mobile video telephony users to communicate with users of the newer packet-based Release 5 mobile video telephony services. Chapter 10, "Supplementary Services and Interworking," outlines approaches to extending mobile video telephony service beyond simple point-to-point communication between two mobile video telephony handsets.

After reading this book you should have a solid technical understanding of the principles behind mobile video telephony.

Migrating to Third-Generation Mobile Networks

Commercial mobile telephony networks have existed for over 20 years. This chapter traces the evolution of mobile telephony since its introduction and the reasons for its development, from the earliest analog networks through to the latest third-generation (3G) systems. The chapter concludes by looking at the ways in which conversational video telephony can be deployed on 3G networks.

1.1 2G Mobile Networks

The first-generation mobile services that were deployed in the 1980s were based on analog techniques. They suffered from relatively poor voice quality and frequent dropouts. A variety of signaling and coding schemes were used, limiting the ability of customers of one mobile network operator to make calls to another, especially between countries. Handsets and network equipment were expensive. As a result they did not become mass-market consumer services but were limited to businesses and other customers who could justify the high cost.

The introduction of standardized second-generation (2G) digital technologies such as *global system for mobile communications* (GSM) brought costs down significantly, making mobile services more affordable, in part due to economies of scale. Technology improvements reduced handset sizes and improved battery life. The introduction of prepay services, combined with the ability to roam across networks and access supplementary services such as voice mailboxes, resulted in rates of growth of customer numbers across the world that were probably unprecedented for any other new product or service. Mobile telephony

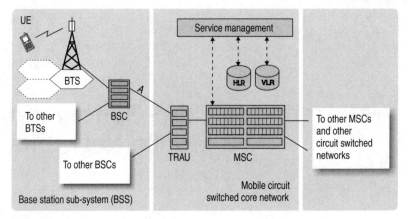

Figure 1.1   GSM 2G mobile network.

is claimed to have grown faster following the introduction of 2G systems than the Internet or television, services that also showed phenomenal rates of growth.

GSM was created by the European Telecommunications Standards Institute (ETSI). It is the dominant world standard for 2G networks, serving over 60% of all mobile customers worldwide. An indication of the impact of widespread standardization can be obtained by comparing the mobile market in Europe with that of the United States, where a range of incompatible 2G mobile technologies are used by different operators. Growth in the number of mobile telephony users in the United States has not been as strong as elsewhere and the ratio of mobile phones to people is still lower than in Europe.

A 2G network based on GSM is shown in Figure 1.1. This includes *base station subsystems* (BSSs) that are connected to the mobile core network. The mobile core network has interfaces allowing it to route calls to customers of other mobile operators and to customers of fixed network operators. Although these are not shown in Figure 1.1, it may also in principle have interfaces to packet networks, allowing its customers to make and receive calls to people using *voice over Internet protocol* (VoIP) on the Internet. In this case gateways would be required, network-based equipment that converts between the signaling protocols and speech coding formats used in the circuit-switched mobile network and those used for VoIP.

1.1.1   The GSM base station subsystem

Figure 1.1 shows that the BSS is composed of *base transceiver station* (BTS) units and *base station controller* (BSC) units. Each BTS covers one or more of the hexagonal cells that provide geographical coverage for mobile customers, and each BSC manages two or more BTS units. The BSC is connected to the mobile core network via a *transcoding and rate adaptation unit* (TRAU). Although the TRAU is not shown in Figure 1.1 as being part of the BSS, its function will also be discussed here.

The BTS provides the air interface allowing digital voice data to be sent to and received from mobile telephones, which are sometimes referred to as *user equipment* (UE). The mobile telephone contains a coder and decoder (codec) that produces a compressed digital speech signal according to one of the GSM Full Rate (GSM-FR), Half Rate (GSM-HR), or Enhanced Full Rate (GSM-EFR) speech coding standards, which result in digital speech signals at 13 kbit/s, 5.6 kbit/s, or 12.2 kbit/s, respectively. Each digital speech call is allocated a 16 kbit/s digital channel between the UE and the BTS.

The BSC manages two or more BTS units. It relays signaling and call data, deals with handover functions when mobile users move between cells, and raises alarms if problems occur. Because the mobile digital voice channels each require 16 kbit/s, four mobile channels can be multiplexed in a single 64 kbit/s channel or PCM timeslot between the BTS and BSC or between the BSC and the TRAU via the circuit orientated A interface shown in Figure 1.1.

The purpose of the TRAU is to convert between the 16 kbit/s digital mobile voice channels used in the BSS and the 64 kbit/s channels used in the mobile core network and in traditional fixed telephone networks. To do this it has to convert or transcode between the compressed GSM speech format and the G.711 *pulse code modulation* (PCM) speech format. If the TRAU is located at the opposite end to the BSC in the link between the BSS and the mobile core network, transmission costs associated with this link are reduced by a factor of 4 compared to locating it at the BSC end of the link.

1.1.2   The 2G mobile core network

The BSC is connected via the TRAU to a *mobile switching center* (MSC) in the mobile core network, which includes a number of MSC units interconnected with one another. Some of these MSCs, known as *gateway MSCs* (GMSCs), have interfaces to other switched networks for signaling, using *signaling system number 7* (SS7), and for transport of voice. The MSC is a traditional switch similar to a *central office switch* in a fixed telephone network.

The other components of the mobile core network shown in Figure 1.1 are the *home location register* (HLR) and *visitor location register* (VLR) databases. The HLR contains information of customer telephone numbers and the VLR that they are currently registered with. The VLR contains information of the customers currently served by the zone it is responsible for. Together with other components that have been consolidated for simplicity into the service management box shown in Figure 1.1, these are responsible for authenticating callers, determining the location of the customer they wish to call, setting up the call, and gathering data for billing purposes. The service management function also provides the *intelligent network* (IN) features that allow services such as rerouting callers to voice mail systems in the event that their desired destination is engaged or unavailable.

The switching infrastructure within the mobile core network for GSM-based 2G systems is based on circuit switching of PCM channels, using switching protocols such as SS7, in very much the same way as in fixed telephone networks. Dedicated 64 kbit/s bearers are set up between customers for the duration of the call. The bandwidth is further limited to 16 kbit/s within the BSS.

1.2   Evolution to 3G

After a sustained period of rapid growth, the mobile operators (and handset and equipment manufacturers) were faced with declining rates of growth as markets began to saturate. As a result they were keen to investigate possible new revenue sources. The success of the *short message service* (SMS), a 2G service allowing users to send and receive text messages up to 160 characters long that exploits existing signaling infrastructure, brought an unexpectedly significant source of revenues to mobile operators. It was a pointer to the future direction in which mobile telephony should evolve—the provision of data services as well as traditional voice telephony.

The data-carrying capacity of 2G mobile networks is limited. At best data can be transferred between the mobile device and the network at 16 kbit/s. In practice using wireless modems the maximum rate that can be achieved is 9.6 kbit/s. This is too low to provide an acceptable user experience. Standards have been produced for enhancements to existing 2G networks to provide some data capabilities based on existing infrastructure. These are usually referred to as 2.5G systems; they are expected in most, but not all, cases to be interim solutions in the run-up to full deployment of 3G.

The vision that drives 3G is that the mobile network will converge with the Internet allowing people to seamlessly access information and to communicate over different networks. High-speed digital services capable of being heavily used by a large number of customers need 3G mobile infrastructure to provide the bandwidth and data carrying capacity.

True 3G networks require additional spectrum allocation. Governments around the world auctioned licenses for this spectrum around 2000.

This raised a great deal of income for the governments concerned, at a high cost to the successful operators. This has delayed the launch of 3G services because the operators did not have sufficient money left over to invest in the required infrastructure. Operators now have more pressure on them to make sure that 3G provides compelling services that produce new revenue streams to recoup their investment in licenses, or to write off the cost of the licenses they bought.

1.2.1 Technology drivers

Since 2G networks were originally launched, a number of improved modulation and signal processing techniques have been developed. These increase the efficiency of spectrum usage at the air interface and provide a potential way of increasing capacity and overcoming the bandwidth and data-carrying limitations of 2G services.

Making use of this increased capacity to offer data services requires that the mobile core network be enhanced. This takes two forms, the upgrading of the existing circuit-switched infrastructure to extend 64 kbit/s channels all the way to the UE, and adding an overlay packet-based core network that operates in parallel with the circuit-switched mobile core to carry data traffic. This requires the functionality of the mobile network service management layer to be extended to support packet-based services. Customers using the packet interface need to be authenticated and billed as reliably as customers using the circuit-switched network. Wherever possible, existing Internet technologies and protocols are being used, and enhanced where necessary to take account of unique features of mobile networks. One example of this is that augmented location and authentication processes are required for data connections to take into account the mobility of customers.

The capabilities of mobile devices need to be extended to offer data services as well as voice services. Developments in this area have increased the computational power of handsets. This enables them to run the required network interfaces, protocol stacks, Internet browsers, and other application software without adversely affecting battery life or weight and therefore portability. Upgraded user interfaces have provided larger screen sizes with improved resolution and integrated cameras to allow video to be captured and displayed. Plug-in cards are becoming commercially available that provide mobile interfaces for laptops and other handheld computers such as *personal digital assistants* (PDAs).

These technology developments have been exploited by standardization groups, such as the *third-generation partnership project* (3GPP) to produce a roadmap and standards for 2.5G and 3G systems.

1.2.2 The evolution path to 3G

Mobile operators have invested large sums of money in their GSM infrastructure. They wish to get a return on this investment, and if possible reuse it where possible in providing 2.5G and 3G services. This desire has informed and impacted the development of the relevant standards, and has led to evolutionary rather than revolutionary approaches to migrating to 3G.

The initial stage in exploiting these technology improvements is the development of 2.5G capabilities to send and receive data, such as *general packet radio service* (GPRS). GSM networks can be upgraded to support GPRS using existing spectrum to provide a packet data interface with data rates that can peak at over 100 kbit/s. This involves upgrading the air interface of existing BTS units and providing BSC units with a packet-oriented interface to an overlay mobile core packet network. The service management layer of the mobile core network needs to be upgraded to provide support for the new services that are possible with GPRS such as the *multimedia messaging service* (MMS).

A further enhancement of GPRS is known as *enhanced data rates for GSM evolution* (EDGE). EDGE introduces a more advanced modulation scheme that allows data rates that are three times those of GPRS without requiring additional spectrum, and is seen as a viable alternative for operators who do not have a license for additional 3G spectrum allocation.

1.3 3G Mobile Networks

The 3G evolution path for GSM is to *universal mobile telephony service* (UMTS). UMTS 3G mobile networks deploy a new *radio access network* (RAN) in place of the BSS of GSM, as well as packet-based infrastructure in the mobile core network.

As we have already seen, the implementation of 3G networks is evolutionary. This process of evolution continues within the deployment of 3G networks, which is broken up into a number of phases. The first of these is usually referred to as Release 99 (after the year in which the recommendations were issued, 1999), though it is also sometimes known as Release 3. The Release 99 architecture includes a circuit-switched component, aimed mainly at voice services and a packet-switched component for data-based services. It is very similar to GSM augmented by GPRS/EDGE, except that the new RAN has greater capacity, provided by wider bandwidth and advanced modulation schemes.

The next phase in the deployment of 3G, known as Release 5, envisages that all traffic including voice will be carried over the packet network, using augmented *Internet protocols* (IPs).

In the short to medium term, however, many mobile operators will have a mixture of legacy GSM, 2.5G, and Release 99 3G networks. The

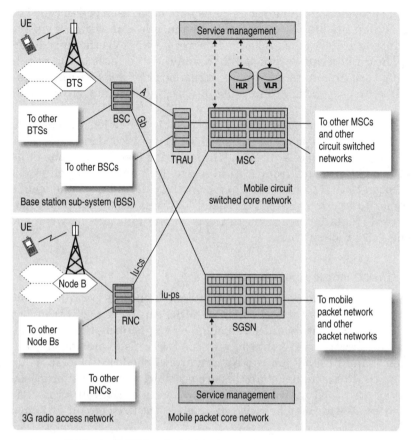

Figure 1.2 Evolved 2G GSM and 3G UMTS network.

network architecture for this setup is illustrated in Figure 1.2, which shows a 2G BSS and a 3G RAN, and a mobile core network split into circuit-switched and packet-switched subnetworks. The connection from the BSS to the packet core network via the Gb interface shown in Figure 1.2 allows 2G customers to access GPRS/EDGE-based services. The RAN connects to the circuit-switched core network via the Iu-cs interface, allowing the mobile operator to reuse 2G infrastructure to provide circuit-switched services to 3G customers.

1.3.1 The radio access network

The RAN shown in Figure 1.2 is made up of Node Bs and *radio network controllers* (RNCs). The Node B is equivalent to the BTS in a GSM BSS, and provides the air interface to the UE. The Node Bs are connected to RNCs, the equivalent of GSM BSCs. Unlike BSCs, the RNCs have direct connections to other RNCs that control Node Bs in adjacent geographical areas to provide soft handover as customers move from cell to cell. The RNC is connected to the circuit-switched mobile core network via the Iu-cs interface and to the packet mobile core network via the Iu-ps interface. The connections within the RAN and the interfaces to the mobile core network are all packet-based, using *asynchronous transfer mode* (ATM) technology, rather than using the circuit-based interfaces of a 2G BSS.

There is no equivalent to the GSM TRAU in the Release 99 3G network. For voice calls, speech is transported in its encoded form using the *GSM adaptive multirate* (GSM-AMR) standard operating in one of its eight encoding modes from 4.75 kbit/s to 12.2 kbit/s. Speech is not transcoded to G.711 and back when being transferred from one mobile customer to another. Speech destined for fixed circuit-switched networks will require conversion to G.711 at the interface between the mobile and fixed networks.

1.3.2   The 3G mobile core network

The MSC in a Release 99 network has equivalent functionality to the MSC in a 2G GSM network—the 2G MSC can be reused through a simple upgrade that includes replacing or augmenting the cards providing the circuit-based A interface with ATM cards to provide an Iu-cs interface, and to send and receive circuit-switched traffic at rates up to 64 kbit/s. Rates higher than this are possible, but for most proposed deployments of 3G networks a maximum rate of 64 kbit/s rate has been adopted—this matches with channels in the fixed circuit-switched network.

Data traffic to and from a UE is routed by the RNC over the Iu-ps interface to an IP switch/router known as a *serving GPRS support node* (SGSN), which is the entry point to the overlay packet-based core network. This performs authentication, allocation of dynamic IP addresses to the UE and IP switching and routing. Once a session is established incoming data traffic can be routed from the SGSN to the next point on the route to its destination via a *gateway GPRS support node* (GGSN), not shown in Figure 1.2. This may be to destinations in the fixed packet network, or to legacy-switched networks via a media gateway.

1.4   Mapping Video Telephony Services onto 3G Networks

In addition to continuing to offer speech telephony, 3G enables mobile operators to increase revenues by offering new services to their customers based on the delivery of digital information at high rates. Video telephony is one such service.

The two possibilities for offering video telephony on Release 99 3G networks are to use the circuit-switched mobile core network to deliver video telephony or to use the packet network.

Significant extra bandwidth is needed to transport video as well as speech. Release 99 3G networks effectively provide a 64 kbit/s circuit-switched path (even if it does make use of ATM between the MSC and the Node B). This is four times as much bandwidth as provided in the BSS of GSM networks, and means that if the speech is encoded at 12.2 kbit/s using the highest mode of GSM-AMR, there is around 50 kbit/s available for video. This is low compared to the bandwidth needed to transport uncompressed video, but video that provides an acceptable user experience can be obtained at this rate with advanced compression techniques. The reader should also bear in mind that the size of the screen that the video will be displayed on is of the order of 10 cm^2, and that head-and-shoulders scenes are likely to be the subject of the majority of mobile video telephone calls.

The alternative to circuit-switched mobile video telephony is to use the packet network. This may provide greater bandwidth, though bandwidth is not guaranteed in the way that it is when a dedicated circuit is available for the duration of the call.

3GPP has recommended packet-based video telephony for Release 5 and later 3G architectures based on the use of the *session initiation protocol* (SIP); however, for Release 99 the recommendation is for circuit-switched mobile video telephony, based on standards derived from the ITU-T H.324 recommendation, "terminal for low bit rate multimedia communication," and commonly referred to as 3G-324M.

The decision to choose circuit-switched 3G-324M is at least in part due to the relative immaturity of the packet core network components and protocols and the associated service management capabilities required. Customers are accustomed to a level of robustness and reliability for speech telephony services that they do not yet demand from the Internet. They will expect video telephony services to be equally robust and reliable. Mobile operators feel comfortable with the well proven and mature systems associated with speech telephony for capturing billing information, circuit-switched video telephony allows them to reuse these same systems. Until later releases of 3G networks when the packet network infrastructure is proven and mature, it seems to make good sense to take the circuit-switched approach.

In places such as Europe, Japan, Korea, Taiwan, and Australia 3G networks are commercially available and operators are offering video telephony services based on the circuit-switched approach. Handsets that support conversational video telephony on 3G networks are being produced by a number of major manufacturers. Japanese operator NTT DoCoMo has announced that its *freedom of mobile multimedia access*

(FOMA) 3G video telephony service, which is based on 3G-324M, passed the milestone of two million customers in January 2004 and three million customers just 2 months later. Perhaps the age of mass-market video telephony has arrived.

1.5 Chapter Summary

In this chapter we have reviewed the development of mobile networks since their introduction. The adoption of standards, advances in technology, the availability of extra spectrum, and the desire of operators to sustain their historical growth rates have driven this development. The latest 3G networks support higher bandwidths for both circuit-switched and packet-switched services, enabling mobile operators to offer a richer set of services.

We have seen that one of these services is mobile video telephony. This can be offered over either the circuit-switched or the packet-switched network. In the future it is likely to make use of the packet-switched network, but for reasons of maturity of technology circuit-switched video telephony services have been launched initially and are likely to continue, at least in the short to medium term.

Chapter 2

Basics of Multimedia Communication

By multimedia communication we mean conversational services that involve the transmission and reception of two or more types of media simultaneously. In this chapter we look at what is involved in conversational multimedia communication and how it differs from simple speech communication. The implications of using different networks—specifically packet networks and circuit-switched networks are described—and the techniques and types of protocols needed to provide an end-to-end capability for multimedia communication are discussed.

This discussion does not focus on any specific multimedia standards but considers the problems inherent in multimedia communication in general terms, and possible approaches to solving them. The aim is to give an understanding that puts into context the multimedia standards used in mobile video telephony that are described in later chapters, and throws some light on the reasons why they work as they do. We will then look at the different components that are required to make up a multimedia user terminal.

The chapter concludes with a review of the need for standards, the key standards making bodies and the most significant standards in the context of mobile video telephony services.

2.1 Features of Multimedia Communication

Multimedia communication is a more complex proposition than more traditional and long-established speech-based communication. To understand this, consider what is involved in making a two-party conversational speech call. To do this a means of sending the speech data in both

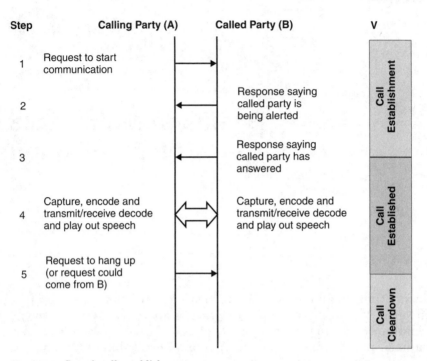

Step	Calling Party (A)	Called Party (B)	V

Figure 2.1 Speech call establishment.

directions must be established. Once this has been put in place, each party can simply begin encoding and sending speech, and decoding received speech. At this high level of abstraction, the process steps are as shown in Figure 2.1. The call is divided into three phases: call establishment, call established, and call clear down.

Compare this with the steps of a similar multimedia call shown in Figure 2.2. In this case an additional phase, multimedia session establishment, is required after the call-establishment phase. Once the call is established the calling multimedia terminal sends a signal or message to the called terminal indicating that it is a multimedia terminal and that it wishes to start a multimedia session. If the called terminal is a compatible multimedia terminal, it will indicate this to the calling terminal by producing an expected response. The two terminals may then negotiate the details of the multimedia session that they will establish. Figure 2.3 is a more detailed breakdown of the multimedia session establishment phase of Figure 2.2, which shows that the terminals exchange capabilities such as the types of multimedia encoding that they support, and agree on a set that they have in common. They then go through a series of negotiations to open the multimedia channels that

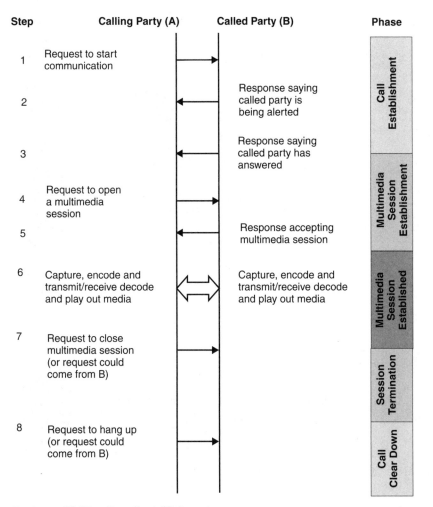

Figure 2.2 Multimedia call establishment.

are needed and the type of encoding to be used. Only when this has been done can media information flow between terminals. As a minimum we would expect to establish one speech channel and one video channel in each direction. More complex sessions may involve multiple video channels and perhaps a channel for exchanging data to allow application sharing. The need to go through this additional phase means that the time taken to establish two-way communication will be longer for a multimedia call than for a speech call. How much longer depends on the complexity of the multimedia session that is being established and the protocols being used.

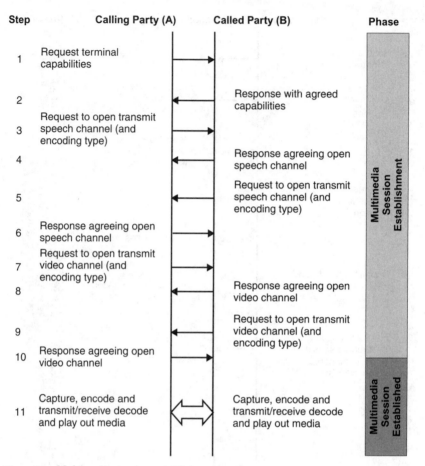

Figure 2.3 Multimedia session establishment.

Multimedia user terminals must capture and digitize the media to be sent. The video and audio is also very likely to be coded in compressed form to reduce the bandwidth required to transmit it, by removing redundant information or information that is not perceptually important to humans. Received media information must be decoded before it can be played out or presented to the user. This means that the multimedia terminal must have a codec for each media format it supports. The subject of media codecs is covered in Chapter 3.

A multimedia terminal must have the capability to send and receive multiple simultaneous media streams and it must support more complex call establishment and control than a simple speech terminal. We will now look at the transmission and call control aspects of a multimedia terminal in more detail.

2.2 Sending and Receiving Control, Media, and Data Information

When two multimedia terminals are communicating over a network, each terminal must be able to unambiguously determine which media stream the information that it receives belongs to, or whether it is data or control information. This is necessary to allow the terminal to process the information correctly. The ways in which this can be done depend on the type of network that is being used for the communication.

The different types of information sent have different degrees of priority. It is critical to making a successful call that control information is sent and received reliably to prevent any mismatch in understanding of the current call status between the terminals. One approach to achieving reliable communication of control information is to ensure that it is transported using a protocol that provides guarantees of delivery, with the capacity to retransmit information that is corrupted or lost. Reliable transport of control information makes session establishment more robust; if the transmission medium is very error prone session setup times may lengthen because of the need to retransmit some control information, but only in severe error conditions will session setup fail completely. The protocols that are most suitable for ensuring reliable transport of control messages are dependent on the network that is being used.

If any of the encoded speech or video (or data) information is lost it does not result in call failure, simply a glitch in the speech or temporary video corruption. While it is not desirable to lose this information, it is not as important to make sure it is reliably delivered. Media streams carry real-time information, so if the capability to retransmit lost or corrupt media information is provided then it must be buffered at the receiver to exploit this capability. Buffering introduces delay, which adversely affects the perceived quality of the call. Minimizing the delay in transmitting media information is as important in conversational multimedia communication as it is in conversational speech communication. As delay increases, the user perception of call quality rapidly reduces, and problems in communication arise, conversation becomes stilted, and long pauses become necessary between sentences to avoid the users at each end talking at the same time.

The delivery of speech information should have a higher priority than video: it is preferable to be able to hear but not see the person that you are communicating with than to be able to see them but not hear them.

2.3 Transmission over Packet Networks

For multimedia terminals using a packet network, a simple approach to allow the receiving terminal to distinguish between different information types is to include information on the intended destination of the information in the address or header part of each packet. This provides an address within the terminal for the process that handles the information. For example, in the case of an IP network like the Internet, different port numbers can be used for the different channels or streams that are established. A predefined port number is needed to initiate a session. Port numbers for the media channels can then be negotiated between the terminals as needed, when they are opened. Terminals based on the H.323 recommendation use port number 1720 for session establishment, and terminals that are based on SIP, which are discussed in Chapter 9, use port number 5060 to initiate calls.

To guarantee that control information is reliably delivered over a packet network, a connection orientated approach at the transport layer can be used, such as the *transmission control protocol* (TCP) for IP networks. TCP has an acknowledgment and retransmission scheme. For example VoIP terminals based on H.323 use TCP/IP for reliable transmission of control information. A perfectly valid alternative approach to achieving guaranteed delivery over an IP network is to use an unreliable transport protocol such as the *user datagram protocol* (UDP) and build reliability in using an acknowledgment and retransmission scheme at the application layer. This approach can be used by SIP.

Video and speech can be transported using connectionless protocols with no delivery guarantees, but with some error checking bits added at the application layer (codec). If UDP/IP is used packets can get dropped when networks are busy, or can arrive out of sequence, so it is useful to provide time stamps and sequence numbers to allow for reordering and detection of lost packets. For IP-based packet networks the *real-time transport protocol* (RTP), which itself uses UDP, provides time stamps and sequence numbers. These can be used to synchronize media streams and to remove the need to send speech information when a speaker is silent. The *real-time transport control protocol* (RTCP) can be used to inform sending terminals about packet loss to allow them to adjust the number of packets being sent where this is feasible, for example by reducing the video encoding rate and quality. Both TCP and UDP include a checksum field in the packet header, allowing detection of corrupt packet headers.

Multimedia communication over packet networks results in an overhead on the payload media data being transmitted, because of the need to include addresses and other packet header information in each packet that is sent. This can be significant, particularly if the payload of the packet is small, as it may be for speech information. Increasing the size of the payload improves bandwidth usage but involves buffering media information, which results in increased end-to-end delay. Conversational multimedia communication over packet networks works best when the available bandwidth is significantly higher than the bandwidth requirements of the media payload.

2.4   Transmission over Circuit-Switched Networks

If the network being used for conversational multimedia communication is a circuit-switched network, it expects to receive a synchronous serial stream of bits from the transmitting terminal. If, for example, the network connection is over a single *integrated services digital network* (ISDN) B channel the terminals must constantly provide 64 kbit/s of data to the network. The same methods of transmitting data that have been described for packet networks could in principle be run over this channel, say by running a TCP/IP stack on top of the *point-to-point protocol* (PPP). The encapsulation of the information within so many nested protocols, providing addressing information that is almost redundant in the context of a point-to-point connection, leads to very inefficient use of bandwidth. The limited residual bandwidth available for payload information is insufficient for multimedia communication of acceptable quality. A more efficient approach is not to send control messages, speech, video, and data information as independent packets but instead to directly combine or multiplex them into a single synchronous serial bit stream, according to a scheme understood by both terminals. This keeps payload overheads to a minimum.

2.4.1   Multiplexing

A typical multiplexer takes information from various sources such as control, video, and speech encoders; and combines it to form a bit stream for transmission to the remote terminal. In the reverse direction a demultiplexer takes in a bit stream, extracts the information within it and passes it to the intended destination—control, video, and speech decoders.

The bit stream produced by the multiplexer must have a recognizable structure, for instance it could be organized into frames. A hypothetical frame structure is shown in Figure 2.4, which shows an 80-octet frame (1 octet is the same as 1 byte; the term octet is used in preference to byte in this context). The frame starts with a single-octet opening flag, which is a unique pattern of bits allowing the receiving terminal to detect the start of a frame by searching for this bit pattern. The bit sequence representing the opening flag must not occur within the frame so measures to prevent it must be taken. The opening flag is followed by a 4-octet header, which may include information on how the remainder of the frame is structured. This tells the receiving terminal how the

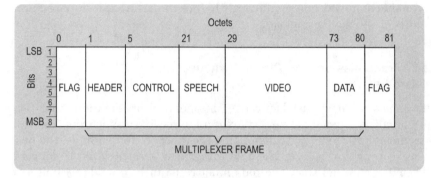

Figure 2.4 A hypothetical multiplexer frame structure.

payload is multiplexed or divided up between control, speech, video, and other data. It may also contain a checksum or *cyclic redundancy check* (CRC) to allow the receiver to confirm that the header information has been received without error. The header may also contain information on the size of the frame. Alternatively, the frame may be fixed length or may be delimited by a closing flag. Making the closing flag identical to the opening flag allows bandwidth to be used more efficiently because the closing flag of one frame could also act as the opening flag of the next frame.

In the example of Figure 2.4 the payload of 76 octets is divided into 16 octets for control messages, 8 octets for speech, 44 octets for video, and 8 octets for data. The frame may be sent over the network starting with either the *least significant bit* (LSB) or the *most significant bit* (MSB) of the first octet. Whichever approach is adopted must be applied consistently so that the receiving terminal knows how to interpret the received data.

With a clear 64 kbit/s channel, multiplex frames like that of Figure 2.4 are equivalent to providing a speech channel of $(64 \times 8/80) = 6.4$ kbit/s between the transmitting and receiving terminals. In a similar way, in the example of Figure 2.4 a 12.8 kbit/s control channel, a 35.2 kbit/s video channel, and a 6.4 kbit/s data channel are provided, assuming that multiplexer frames are formed and sent at the maximum rate possible and that the payload is multiplexed in the same way in each frame.

It may not always be possible to form and send multiplexer frames conforming to this structure. There are times during the call, particularly during the early stages before the terminals have started to exchange media, when there is no video or speech information available. If this is the case, a strategy that can be adopted is to form a multiplexer frame that is up to 20 octets long, consisting of a 4-octet header and up to 16 octets of control information. This strategy assumes that whilst the general structure of a multiplexer frame is as shown in Figure 2.4, frames that represent partial versions of this can be sent, with the end of the frame able to occur at any point, indicated by a closing flag.

There may also be times when there is not even any control information available to send, perhaps because the transmitting terminal is awaiting a response from the remote terminal. When there is no frame ready to send, the terminal must continue to send a constant bit stream over the network. It can do this by sending stuffing flags, which at their simplest could comprise repeated transmissions of the opening flag. An alternative is to repeatedly send the opening flag followed by a frame consisting only of a header.

Having just a single multiplex frame structure or template can be restrictive and inefficient—there may or may not be information of a particular type to send. Speech information is relatively constant, whereas the amount of video information tends to vary with time. Having a number of alternative multiplexer frame structures that can be used provides the terminal with a way to cope efficiently with statistical fluctuations in the different streams that are being multiplexed. For example there could be structures available where the entire payload was devoted to one media type, or only to control messages. The definition of a number of multiplex structures could be provided in several ways. The first of these is to fully describe the structure of the frame in the header of each frame, this provides maximum flexibility, but may lead to bandwidth inefficiencies because the size of the header required may be large. A second approach is to have a number of predefined frame structures, which are known to each terminal. In this case an N bit multiplex code in the header of the frame can be used to signal which of 2^N predefined multiplex structures is being used. A third compromise approach that combines the flexibility of the first approach with the efficiency of the second is to have just one predefined frame structure, whose payload is dedicated entirely to control information used in the session establishment phase. Definitions of the other multiplex structures that will be used during the session are exchanged by the terminals during multimedia session establishment, along with the multiplex codes that will be used to reference which one is being used. As we will see in later chapters, this is the approach used in 3G-324M terminals.

2.4.2 Detecting multiplexer frames

The first action that the demultiplexer has to perform on the stream of bits that it is receiving is to detect the frames within the stream and extract them. It does this by detecting and throwing away stuffing flags and detecting the opening and closing flags of the frames.

If the bit sequence representing the opening (and closing) flag occurs within the frame, an error will occur. There are a number of approaches to minimizing or preventing the occurrence of spurious flags. One approach is for the transmitter to detect them before sending and take steps to prevent them. In a technique known as bit stuffing, assuming the opening flag is 7E in hexadecimal (0111 1110 in binary) the transmitter can examine the octets in the frame between the opening and closing flags and insert a binary 0, if it finds five binary 1s in a row. After detecting frame boundaries the receiver throws away any 0 that follows five 1s. The detection of frame boundaries can be performed simply by searching for the flag sequence in the incoming bit stream. In the remainder of this book, hexadecimal numbers will generally be represented by preceding them with 0x; thus hexadecimal 7E is represented as 0x7E.

If the network being used has an octet aligned structure, and the multiplexer frame is also octet aligned (i.e., all fields within the frame start on octet boundaries) then it is very beneficial to ensure that the octets of the multiplexer frame align with the octets of the underlying network. One benefit is that it reduces the probability of detecting flag emulation; binary sequences that match the opening flag but are not octet aligned can be ignored. The scheme of inserting zeros to avoid flag emulation has the drawback that it destroys the alignment of multiplexer octets with the octets of the underlying network.

Alternative approaches are possible at the octet level. For example if a flag in the payload of the multiplexer frame is detected by the transmitter it could insert an octet of a specified value before it (say 0x7D), and one or more bits of the 0x7E are inverted or complemented, for example the 6th bit, giving 0x5E. So if 0x7E were to occur in the payload it would be represented by 0x5E. If the receiving terminal detects the value 0x7D, it can check the octet that follows; and if it is 0x5E, it can discard the 0x7D and reinvert the 6th bit of the following octet to restore it to 0x7E. If the following value is not 0x5E, it will pass 0x7D and the following value on. There is still a chance that a spurious 0x7D 5E will occur in the frame, and this will get corrupted; however, the probability of this happening is much lower than the probability of flag emulation if no scheme is in place to avoid it.

A further possible approach is to have an opening flag that is two or four octets instead of 1 octet. This reduces significantly the probability that there will be an emulation of the flag within the frame. The use of correlation to detect the flags (rather than matching) and comparing the correlation output to a threshold provides increased resilience in an error-prone environment.

Including the frame length in the header information is a further way to improve robustness; the closing flag, which may be the opening flag of the next frame, should only be detected where indicated by the frame length value in the header.

2.4.3 Error protection

Speech transmission is likely to consist of sending regular small chunks of information, whereas blocks of video information may be more variable in size and frequency of transmission—so the impact of transmission errors may affect speech and video differently. For this reason appropriate approaches to error protection may vary depending on the media type being transmitted, so schemes are best applied on a per stream basis rather than to the multiplexer frame as a whole.

Before information from a particular source is inserted into the multiplexer frame it may need error protection added to it, and/or sequence numbers added, to provide additional protection and robustness. The extent and exact nature of this is dependent on the media type and the characteristics of the channel. Some error-protection measures may be implemented by the source codec supplying the bit stream to the multiplexer. If additional protection is required it can be performed in an adaptation layer, giving the multiplexer a two-layer structure: the upper adaptation layer that applies media specific error protection and the lower layer that forms the multiplexer frames for transmission.

Common approaches to error detection include the addition of checksums or CRC bits to the information. A checksum can be formed by adding all the octets in the information to be transmitted. This sum can then be sent along with the information. The receiving terminal independently computes the checksum and compares it with the checksum sent—if they match the data is highly likely to have been sent without error.

CRCs are generally computed by using the information to be checked as a single binary number to form a numerator, dividing it by another binary number (the denominator) and using the remainder as the CRC. The CRC is then sent along with the information. The receiving terminal can independently compute the CRC and compare it with the CRC that has been sent, if they match the information has been transmitted without error.

CRCs are often described in terms of polynomials. Taking a simple example from the ITU-T H.223 recommendation, a 3-bit *header error control* (HEC) field is a CRC computed for the 4-bit *multiplex code* (MC) field. The HEC field is specified to be the remainder obtained by reversing the bit order of the content of the MC field, multiplying it by x^3 and dividing it modulo 2 by the generator polynomial $P(x) = x^3 + x + 1$. The resulting HEC field value is constructed by reversing the order of the remainder bits.

MC value   = 0101
Product   = 1010000
Generator poly = 1011


```
  1011  │1010000 │ Don't care
         1011
         0010
         0100
         1000
         1011
Remainder =  011
```

Figure 2.5 CRC computation example.

HEC field = 110

Computing the HEC value for a particular MC is simpler than it sounds from the above description. The generator polynomial P(x) and x^3 are binary numbers, formed by setting x = 2. This gives P(x) = 1011 and x^3 = 1000 in binary. If we take as an example the MC value 0101, we form the denominator by reversing the order of the bits and multiplying by 1000 to give 1010000. This is then divided modulo 2 by 1011.

Modulo 2 division is performed by performing a bit-wise exclusive OR (XOR) operation. This is shown as a long division operation in Figure 2.5. If the MSB of the denominator is 0, it is simply discarded and the number is shifted left until there is a 1 in the most significant position. The top four bits are then XORed with the generator polynomial. The most significant bit of the result is discarded and if there are any remaining trailing digits one is brought down to form the LSB of a 4-bit word. If the MSB of this word is zero, it is discarded, the word is shifted left, and another digit is brought down. If the MSB is a 1, it is XORed with the generator polynomial and the MSB of the result is discarded. This process continues until there are no more trailing digits left, the result of the final operation is the remainder. In our example this is 011, so for an MC field with the value 0101, the value of the associated CRC in the HEC field is 110.

This is a very simple example of computing a CRC; however, it is sufficient to understand the principles behind computing any CRC. CRCs may be computed for much larger fields than the 4-bit MC field, for example they may comprise many bytes of speech or video information, and the CRC calculation may involve more sophisticated pre- and postconditioning (ways of forming the numerator and the denominator, and deriving the CRC from the result).

2.4.4 Guaranteeing delivery of control information

It is generally considered that circuit-switched networks provide better guarantees of delivery than packet-switched networks. One notable exception to this is mobile or wireless switched networks, where occasional loss of signal on the air interface occurs due to factors such as buildings obscuring the transmitter or the terminal operating on the edge of a cell. This results in bursty patterns of error or data loss. Just as for packet networks, it is vital that control information is received reliably. This can be achieved by using an acknowledgment protocol, where the receiving terminal sends an acknowledgment (Ack) message to the transmitting terminal to confirm that it has received a control message. If the transmitting terminal does not receive an Ack message within a defined period of time, it will retransmit the control message again. It will continue to do this until it receives an Ack or until the maximum number of allowable retries has been exceeded, at which point it will typically terminate the call gracefully.

One approach to providing this acknowledgment is shown in Figure 2.6 where a control acknowledgment protocol is implemented as a process

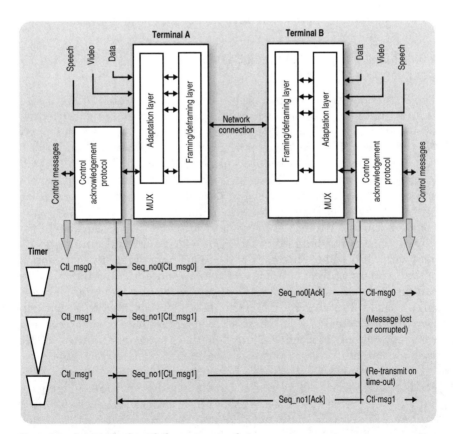

Figure 2.6 A control acknowledgment protocol.

between the source of control messages and the multiplexer. Control messages are encapsulated in control acknowledgment protocol messages. Referring to Figure 2.6, the first control message (Ctl_msg0) has a sequence number added to it by the control acknowledgment protocol. It is then transmitted and an acknowledgment is generated by the peer control acknowledgment protocol process and received successfully within the allowed time period. The second control message (Ctl_msg1) is not acknowledged before the timer expires. The control acknowledgment protocol process on the transmitting terminal retransmits Ctl_msg1 and this time successfully receives an Ack message. The source of control messages is only informed that the message has been successfully delivered after the control acknowledgment process has received the Ack. The inclusion of sequence numbers in the Ack allows the control acknowledgment protocol to associate Ack messages with specific outgoing encapsulated control messages. This prevents confusion arising in circumstances where the receiving terminal sends an Ack after the timer has expired and the transmitting terminal has retransmitted the control message.

As we will see, an approach similar to this is used in 3G-324M.

2.5   Establishing a Multimedia Session: Command and Control

The simple description of a speech call that is illustrated in Figure 2.1 is incomplete—it assumes successful call completion and doesn't show what happens when the called party is unavailable, or the called party fails to answer. If there was no formal, designed-in way to handle such circumstances, the caller could in principle hang on indefinitely waiting for an answer, and in doing so tie up valuable network equipment and resources without the operator deriving any revenue. If the called party is busy or fails to answer, the calling party may be expected to abandon the call attempt and hang up; however, the system should not rely on this. A mechanism must be provided that will automatically deal with all possible outcomes of a call attempt. To do this the notion of time, timers, and call state need to be introduced. The calling process can be modeled as two communicating *finite state machines* (FSMs), one on each side of the call. For a simple *general switched telephone network* (GSTN) telephone call these state machines may reside in the network equipment. For multimedia terminals some or all of the FSMs tracking call and session status will reside in the user terminals.

For a multimedia call, once a bearer or connection is in place, the FSMs needed to establish a multimedia session must be designed to handle all possible outcomes. We will refer to these FSMs as *signaling entities* (SEs). A multimedia terminal may comprise multiple SEs—each handling a different process. Referring to Figure 2.3, there may be an

SE that is responsible for establishing a common set of capabilities and other SEs responsible for opening media channels. These SEs communicate with peer SEs in the remote terminal using messages with syntax and semantics that are understood by both sides.

To understand how SEs and message formats can be specified we will use an artificial example—a terminal that is capable of establishing unidirectional speech channels for speech encoded either as G.723.1 (a popular format for VoIP terminals) or as GSM-FR (a format used in GSM mobile networks). Up to one speech channel can be established in each direction, and they need not be of the same encoding type. In our example, if a terminal fails to get a response from the remote terminal it will retry up to five times before informing the SE user that its request to open a channel has failed. The channel can only be closed by the terminal that opened it.

We will assume that it is a terminal for a circuit-switched network. There are two types of multiplexer frame available: one carries only speech data and the other carries only control information. Once the bearer is established, the multiplexer sends stuffing flags until the user asks to open a speech channel of a particular type. This speech channel is a virtual or logical channel within the multiplexer frames. Once the speech channel is opened multiplex frames containing this information can be constructed and transmitted. In this interaction the first few multiplex frames transmitted and received will be control information to establish the speech channel, after that multiplex frames containing encoded speech will be transmitted until the user asks to close the channel.

2.5.1 Describing signaling entities

We will adopt a model for the control part of the terminal in which there will be an outgoing SE and an incoming SE. The outgoing SE communicates with a peer incoming SE in the remote terminal, and the incoming SE communicates with a peer outgoing SE in the remote terminal by sending and receiving control messages. The incoming and outgoing SEs within a single terminal are independent of one another, because the establishment of channels in each direction is independent.

The SE communicates with an SE user through signals called primitives. Note that the SE user is not the person making the call, but part of an application layer that provides overall management of the terminal and communicates with a *user interface* (UI). The UI may allow the user of the terminal to enter numbers on a keypad, and may display status messages provided by the SE.

Each SE has an input queue that can contain signals from three sources: messages from the peer SE, primitives from the SE user, and

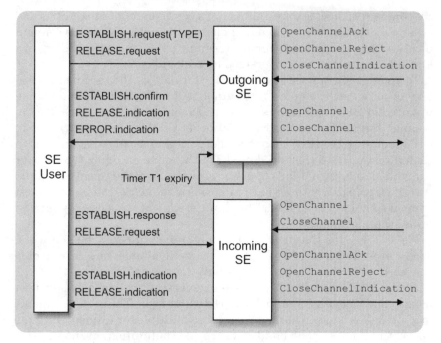

Figure 2.7   The SE/SE user model.

events generated internally to the SE such as expiry of an internal timer. These signals are acted upon by the SE in the order that they are placed on its input queue.

Figure 2.7 shows the incoming and outgoing SEs, the set of messages they can send to each other and the primitives that can be exchanged between the SE user and the SEs. The outgoing SE requires a timer T1, which it can set (start) and reset (stop). The timer is used to set a maximum period that the outgoing SE will wait for a response from its peer incoming SE. Once set, if it is not reset within a fixed time, it will expire, generating an event that is placed on the input queue of the SE. It will also have an internal counter sv_COUNT, which is initialized with the value 1 when a primitive is received from the SE user requesting that a channel of a particular type be opened, and is incremented every time the channel fails to be established because the peer terminal fails to respond before the timer T1 expires. When sv_COUNT reaches a value N100, which will have a value of 5 in this example, it will abandon the attempt to establish a channel.

The outgoing SE can be in one of three states: IDLE when no channel is established, AWAITING ESTABLISHMENT (AE) when channel establishment is in progress but not completed, ESTABLISHED (EST) when the channel is established. The incoming SE has three states that are identical to those of the outgoing SE. The SE changes from one state to another depending on the message, primitive or event that it retrieves from its input queue.

Each SE is initially in the IDLE state. To establish a channel the SE user uses the **ESTABLISH.request(TYPE)** primitive to instruct the outgoing SE to send an OpenChannel message, with the parameter TYPE specifying whether the channel is to be G.723.1 or GSM-AMR. The outgoing SE sends the OpenChannel request to its peer incoming SE on the remote terminal.

The actions and state transitions of the outgoing SE in response to the messages, primitives, and events it receives when in each possible state are shown in Figure 2.8. Figure 2.9 shows the same information for the incoming SE. In these figures *send* is used to indicate that a message is sent to the peer SE, and *inform* is used to indicate that a primitive is communicated to the SE user.

It can be seen from Figures. 2.8 and 2.9 that the SE has to handle all conditions that might arise when it is in any given state. For example, when the outgoing SE is in the AWAITING ESTABLISHMENT state it has to have a defined action to cover the possibilities of receiving a message confirming that its request has been successful (OpenChannelAck), receiving a message rejecting its request (OpenChannelReject), not receiving a response within the allowed time (Timer T1 expiry), or the user changing their mind and deciding not to open a channel (**RELEASE.request**). When the outgoing SE is in the IDLE state it must have explicit actions in case an OpenChannelAck or OpenChannelReject is received, which could happen if the peer incoming SE is taking too long to provide a response. The incoming SE has to deal with such circumstances as receiving an OpenChannel request when it is in the ESTABLISHED state and already has a channel open. The behavior specified in Figure 2.9 is to treat this as a request to close the existing channel and open another one, which may possibly specify a different encoding type.

2.5.2 Specification and description language

A way of graphically representing SEs is to use the *specification and description language* (SDL) format. SDL is a notation for describing systems in terms of their structure, their internal and external communication, and their behavior as a set of interacting state machines. SDL has been standardized by ITU-T in recommendation Z.100. Figure 2.10 shows some of the symbols used in SDL: there are symbols representing the state of the SE; symbols indicating the sending and receipt of messages, events, and primitives; symbols for internal tasks or computations; and decision symbols allowing conditional branching. SDL has many more

Current State=IDLE	
Message, Primitive, or Event Received	Actions
ESTABLISH.request	Increment sv_COUNT; Send OpenChannel Request; Set Timer T1; Move to State AE
OpenChannelAck	Do nothing
OpenChannelReject	Do nothing

Current State=AWAITING ESTABLISHMENT (AE)	
Message, Primitive, or Event Received	Actions
OpenChannelAck	Reset Timer T1; Inform ESTABLISH.confirm; Move to State EST
OpenChannelReject	Reset Timer T1; Inform RELEASE.indication; Move to State IDLE
Timer T1 Expiry	If sv_COUNT<N100 { Inform ERROR.indication; Send CloseChannel Command; Inform RELEASE.indication; Move to State IDLE } Else { Sv_COUNT:=sv_COUNT+1; Send OpenChannel Request; Set Timer T1 }

Current State=ESTABLISHED (EST)	
Message, Primitive, or Event Received	Actions
RELEASE.request	Send CloseChannel Command; Move to State IDLE

Figure 2.8 Outgoing SE operation.

symbols, but the set shown in Figure 2.10 are sufficient to describe our example of command and control for a speech channel.

The SDL representations for the outgoing and incoming SEs are shown in Figures. 2.11 and 2.12 respectively. There is one SDL diagram for each state that the SE can be in. There is one branch within each SDL diagram for each of the possible messages that can appear on the input queue of the SE when it is in this state. Each branch ends in a destination state (which may be the same as the starting state).

Current State=IDLE	
Message, Primitive, or Event Received	Actions
OpenChannel	Inform ESTABLISH.indication; Move to State AE
CloseChannel	Do nothing

Current State=AWAITING ESTABLISHMENT (AE)	
Message, Primitive, or Event Received	Actions
ESTABLISH.response	Send OpenChannelAck; Move to State EST
RELEASE.request	Send OpenChannelReject; Move to State IDLE
OpenChannel	Inform RELEASE.indication; Inform ESTABLISH.indication
CloseChannel	Inform RELEASE.indication; Send CloseChannelIndication; Move to State IDLE

Current State=ESTABLISHED (EST)	
Message, Primitive, or Event Received	Actions
OpenChannel	Inform RELEASE.indication; Inform ESTABLISH.indication; Move to State AE
CloseChannel	Inform RELEASE.indication; Send CloseChannelIndication; Move to State IDLE

Figure 2.9   Incoming SE operation.

The graphical representations of Figures. 2.11 and 2.12 represent the same information as is contained in Figures. 2.8 and 2.9. They may make it easier to visualize what is happening than the descriptions in Figures. 2.8 and 2.9 do, but they are large, even for our simple example. SDL is used to describe command and control for multimedia standards such as 3G-324M.

2.5.3   Message syntax, semantics, and representation: ASN.1 and PER

The messages between the SEs in our example can be classified as falling into four different types: requests, responses, commands, and indications. OpenChannel is a request message that requires the receiving SE to take some action and provide a response. OpenChannelAck

Symbol	Description	Example

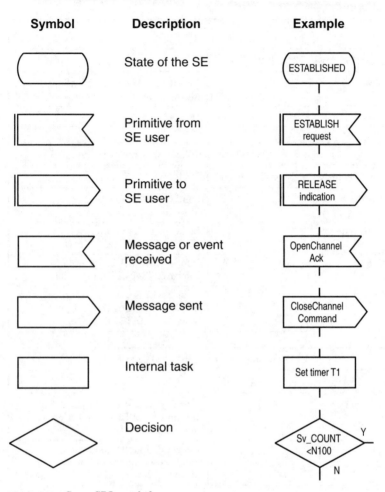

Figure 2.10   Some SDL symbols.

and OpenChannelReject are responses, generated in response to a request message. CloseChannel is a command message that requires the receiving SE to take some action but does not require a response. CloseChannelIndication is an indication message, which does not require the receiving SE to do anything.

We have not yet discussed how the messages exchanged between SEs can be specified. *Abstract syntax notation 1* (ASN.1), described in the X.680–X.683 recommendations of ITU-T provides a formal way of specifying data structures, which are independent of any specific programming language or implementation, so that these data structures can be sent between communicating processes. It resembles a programming language

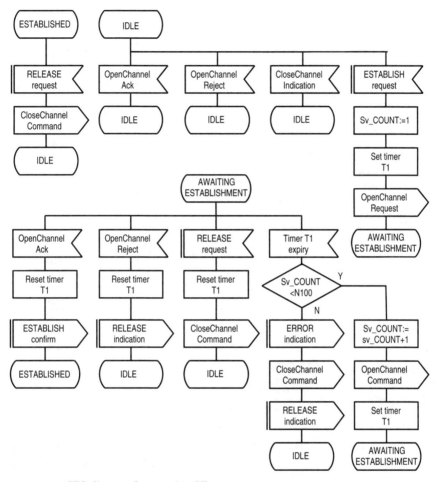

Figure 2.11 SDL diagram for outgoing SE.

that only comprises type declarations—it has no operators for manipulating the data.

ASN.1 can be used to describe the message set for communications protocols including control protocols—it is not a protocol in itself, simply a tool for describing such protocols.

Figure 2.13 shows an ASN.1 like specification for the syntax of the messages in the control protocol we have invented in our example.

Lines beginning with a double dash (– –) in Figure 2.13 are comments. Using the syntax described in Figure 2.13, the format of the messages that would be exchanged between an outgoing and an incoming SE to successfully open a G.723.1 speech channel at the first attempt and subsequently close it are shown in Figure 2.14.

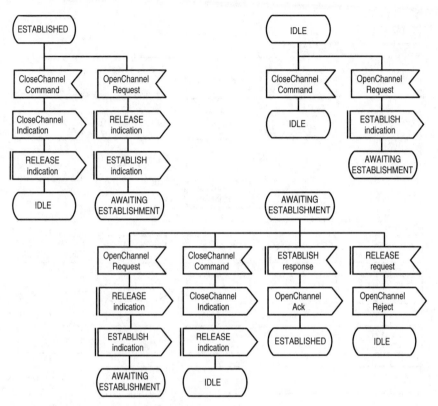

Figure 2.12   SDL diagram for incoming SE.

The messages specified in Figure 2.13 could be encoded as *American standard code for information interchange* (ASCII) text strings and sent in that format. This is inefficient particularly for complex data structures describing real multimedia terminals, where messages such as those describing terminal capabilities are potentially large. An attraction of ASN.1 is that ways of encoding ASN.1 structures into compact binary representations have been developed. The use of *basic encoding rules* (BER), specified in ITU-T recommendation X.690, is one such method. BER is a very general way of encoding ASN.1 messages, which uses a *tag, length, value* format, where *tag* is a binary code representing the name of the structure, *length* is the number of bytes or octets that are required to hold the value, and *value* contains the data in the structure. The *value* can itself be in the form (*tag, length, value*) allowing nesting to any depth. The use of BER leads to efficient binary representations of ASN.1 structures.

The *packed encoding rules* (PER), specified in ITU-T recommendation X.691 produce an even more compact representation of the ASN.1 structures, by removing unnecessary or redundant information.

```
--SYNTAX OF SE MESSAGES

ChannelControlMessage        ::=CHOICE{
       request               RequestMessage
       response              ResponseMessage
       command               CommandMessage
       indication            IndicationMessage
}

RequestMessage       ::=CHOICE{
openChannel          OpenChannel
}

ResponseMessage      ::=CHOICE{
openChannelAck               OpenChannelAck
openChannelReject            OpenChannelReject
}

CommandMessage       ::=CHOICE{
closeChannel                 CloseChannel
}

IndicationMessage ::=CHOICE{
closeChannelIndication     CloseChannelIndication
}

--Channel Signalling Definitions

OpenChannel ::=SEQUENCE{
dataType                     Datatype
}

DataType      CHOICE{
       g7231         NULL
       gsmfr         NULL
}

OpenChannelAck       ::=SEQUENCE{}
--this message has no parameters

OpenChannelReject ::=SEQUENCE{
       cause CHOICE{
                 unspecified  NULL
       }
}
CloseChannel         ::=SEQUENCE{}
--this message has no parameters

CloseChannelIndication        ::=SEQUENCE{}
--this message has no parameters
```

Figure 2.13   Specification of messages for speech channel example.

For example, if the length of the *value* is fixed for a particular structure, then the length is not encoded.

The use of ASN.1 to describe a set of messages for a control protocol and the encoding of these messages using PER enables control messages

```
--OUTGOING SE REQUEST

Request OpenChannel{
       Datatype = g7231 <<null>>
}

--PEER INCOMING SE RESPONSE

Response OpenChannelAck{}

--The channel is now open and data can be sent

--OUTGOING SE COMMAND

Command CloseChannel{}

--PEER INCOMING SE INDICATION

Indication CloseChannelIndication{}

--The channel is now closed
```

Figure 2.14   Messages to open and close a speech channel.

to be transmitted over a network in a very compact form. This is valuable both in reducing the bandwidth requirement and in reducing the susceptibility of the control messages to errors—the shorter the length of a control message, the less likely it is to be corrupted on a network with a given bit error rate.

2.6   Components of a Multimedia User Terminal

Having discussed the key aspects of multimedia communication, we can now review the components that are required in a multimedia user terminal for circuit-switched networks. Figure 2.15 is a schematic of a possible multimedia user terminal for a circuit-switched network. It is made up of three main groups of functions, which are characterized by whether or not they are within the scope of internationally recognized standards, and for those functions that are whether or not they are within the scope of multimedia standards. We will now consider each of these groups in turn.

Referring to Figure 2.15, the group of functions that fall within the scope of multimedia standards comprises the speech and video codecs, the control signaling entities, the control acknowledgment protocol, and the multiplexer, which applies media specific error protection via an upper adaptation layer and forms frames of information for transmission

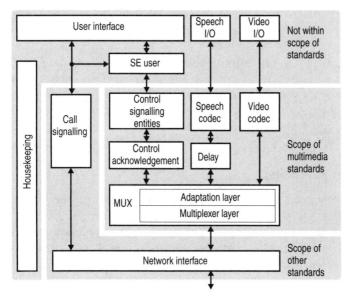

Figure 2.15   Components of a circuit-switched multimedia terminal.

in a lower multiplexer layer. A delay element is also included, which allows any difference in the latency associated with video and audio streams to be equalized to provide "lip synch."

The second group of functions relates to the network interface and the call signaling required to establish the bearer. These functions are specified by standards relating to the specific network in use: for example if ISDN is used the call signaling may be as specified in ITU-T recommendation Q.931. If the network needs to be multimedia aware, for the purposes of routing or billing the call appropriately or setting up a bearer with the correct characteristics, the call signaling may need to reflect this. For instance it may need to signal that a transparent channel is required.

The last group of functions comprises the user interface to allow the end user to interact with the terminal, the speech and video *input/output* (I/O), the SE user process and the housekeeping process, responsible for initialization of the system at start-up, power management, and other miscellaneous functions. It is in this area that designers have the greatest freedom to differentiate their product from competing products through such things as provision of more compelling user interfaces, higher resolution displays, low noise cameras and pre- and postprocessing of the video and speech signals. Information on the capabilities of the terminal resides within the SE user function. It could be argued that the SE user function is part of the system that falls within the scope of multimedia standards; however, this only applies to the primitives and interface to the control signaling entities, and the sequence in which the SEs are initiated.

The components of a multimedia terminal intended for use on a packet network such as the Internet are broadly similar, except that the multiplexer is not required and is replaced by a packetization function for each information stream—the guaranteed delivery and error protection mechanisms provided by protocols at the transport layer are used.

2.7   Multimedia Standards

Standards are essential if multimedia communication is to be widely adopted. Standards allow products from different manufacturers to interwork with one another. Proprietary equipment leads to closed user groups with little or no communication between communities; standards allow open communication and remove barriers that would otherwise hinder the widespread take-up and growth of multimedia services.

A number of standards bodies are active in the multimedia area. Cooperation between these groups has greatly increased in the last few years. Standards tend to have been developed for specific networks, with varying degrees of commonality of standards across networks. This looks increasingly problematic as the emphasis in network development has moved to convergence and universal access to a common set of services over diverse networks. This problem will be discussed in Chapter 10.

We have seen that a multimedia terminal is composed of a set of disparate building blocks. Similarly, multimedia standards are not standalone but reference a set of constituent standards that separately specify components such as the codecs, multiplexing, and control protocols. These, together with the top level standards document represent a complete specification of the standard.

2.7.1   Standards making bodies

Probably the first body to formulate standards for conversational multimedia was the International Telecommunications Union (ITU). Network operators are strongly represented among its membership, which also includes equipment manufacturers and others. The International Standards Organization (ISO) is another group that involves itself in multimedia standardization. The Internet Engineering Task Force (IETF) concerns itself with multimedia systems that are intended for use on IP-based networks.

The ITU, ISO, and IETF are globally recognized standardization bodies. Different regions of the world have more localized standardization groups such as the ETSI, the Association of Radio Industries and Businesses (ARIB) and Telecommunications Technology Committee (TTC) in Japan, the China Communications Standards Association (CCSA), the Korean Telecommunication Technology Association (TTA), and the Telecommunications Industry Association (TIA) and the Alliance for Telecommunications Industry Solutions (ATIS) bodies in the United States.

There are increasing links between all these standards bodies, with standards set by one body referencing standards produced by others. More formal cooperation between the various groups is evidenced in organizations such as the 3GPP, which brings together ARIB, CCSA, ETSI, ATIS, TTA, and TTC. The 3GPP body produces specifications for 3G systems based on the UMTS standard and the evolving GSM network.

A related group 3GPP2 performs a similar function to 3GPP for 3G systems based on CDMA2000 and has ARIB, CCSA, TIA, TTA, and TTC as its partners.

The aims of 3GPP and 3GPP2 are to produce technical specifications covering all aspects of 3G mobile networks—from the terminal equipment and radio interface to the network infrastructure, service management, and services, including video telephony.

2.7.2 ITU-T standards

The ITU organization is divided into two standardization sectors, the *telecommunication standardization sector* (ITU-T) and the *radiocommunication sector* (ITU-R). The standards produced by the ITU are known as recommendations. ITU-R is involved in aspects of mobile video telephony relating to bandwidth allocation and description of the air interface. ITU-T has a wide range of telecommunications standardization activities. Its H series of recommendations address audiovisual and multimedia systems.

The H series of recommendations are further subdivided by number range. Recommendations in the range H.300–H.399 are reserved for "systems and terminal equipment for audiovisual services," these represent high level specifications of complete systems and terminals aimed at different networks. H.220–H.229, "transmission multiplexing and synchronization" deals with transmission aspects and H.240–H.259 "communication procedures" covers control protocols. H.260–H.279 "coding of moving video" covers video coding recommendations. Other number ranges cover different aspects of multimedia services, such as encryption and supplementary services.

ITU audio coding standards are covered in the G series of recommendations "transmission systems and media, digital systems, and networks".

Other recommendations that the H series may reference are the V series "data communication over the telephone network" for modem-based communication; the T series "terminals for telematic services" for data transfer; the X series "data networks and open system communications," which includes ASN.1 and PER; and the Z series "languages and general software aspects for telecommunication systems," which covers SDL.

An early multimedia recommendation produced by ITU-T was recommendation H.320, ratified in 1990, which was aimed at multimedia systems operating over one or more ISDN 64 kbit/s links—the development work that led to this standard took place almost 20 years ago. H.320-based terminals have never resulted in mass market consumer products but they have achieved some popularity for videoconferencing among business users, mainly in Europe.

Following H.320, further recommendations that were developed included H.323 "packet-based multimedia communications systems" for multimedia communications over *local area networks* (LANs) and the Internet, and H.324 "terminal for low bit-rate multimedia communication" for terminals operating over the GSTN using modems. H.323 and H.324 have some commonality. Both use H.245 "control protocol for multimedia communication" to establish multimedia sessions and H.261 "video codec for audiovisual services at p × 64 kbit/s" as the mandatory video codec. They differ in that H.323 uses H.225.0 for call signaling and to send media data using UDP/RTP, where H.324 uses V.250 for call setup and H.223 "multiplexing protocol for low bit rate multimedia communication" to multiplex media and control information for sending over a synchronous modem interface conforming to V.34 and V.8.

H.323 and H.324 were first ratified in 1996. Although it was originally proposed for multimedia terminals H.323 became most popular as a protocol for VoIP terminals, which did not make use of its support for video. H.323 video terminals and PC-based software implementations have been developed, of which Microsoft Netmeeting is probably the most well known. H.324 failed to achieve widespread success as a standard for GSTN-based video telephony—one of the main reasons for this is that the available bandwidth over a GSTN modem is simply insufficient to provide an acceptable user experience.

Despite the limited success of H.324, standardization activities continued, with the aim of broadening the applicability of H.324 to cover multimedia over mobile networks. This resulted in Annex C of H.324, "multimedia telephone terminals over error prone channels" and the related annexes A, B, C, and D of H.223, extending the multiplexing protocol to apply to channels with increasing susceptibility to errors. Terminals that implement H.324 with Annex C have come to be known as H.324M terminals, where the M refers to mobile.

2.7.3   Relevant IETF standards

The IETF produces standards in the form of *requests for comment* (RFCs). The key RFCs for Internet-based transport are IP (RFCs 791, 919, 950, 922), TCP (RFC 793), and UDP (RFC 768). These are used by ITU-T recommendation H.225.0, which is a component of H.323, along with the RTP of RFC 1889 and the RTCP of RFC 1890.

The IETF is also responsible for the specification of *Internet protocol version 6* (IPv6) through a series of RFCs, which extend the range of Internet addresses. Internet addresses based on *Internet protocol version 4* (IPv4) are in very short supply—deployment of IPv6 is necessary for later, packet-based releases of 3G networks.

Finally the IETF has developed SIP, specified in RFC 3261, and the *session description protocol* (SDP) of RFC 2327. These provide an alternative approach to H.323 for IP-based multimedia communications, which has been adopted by 3GPP for later releases of 3G networks. This approach is discussed in Chapter 9.

2.7.4   ISO standards

The ISO is responsible for the *moving picture experts group* (MPEG) series of standards. The MPEG4 standard is of most relevance to conversational multimedia at rates below a few hundred kilobits. MPEG4 video is specified in the ISO/IEC 14496-2 standard.

The specifications of ASN.1 and PER are joint ITU/ISO standards.

2.7.5   3GPP and 3GGP2

The 3GPP and 3GPP2 groups are made up of organizational partners representing other standards making bodies, as well as individual members. 3GPP is organized into *technical specification groups* (TSGs), which produce technical specifications in areas such as the radio access network, the core network, terminals and services, and systems aspects for 3G networks based on an evolved GSM core. Since 2000 3GPP has also taken responsibility for the maintenance and evolution of GSM standards from the *special mobile group* (SMG) within ETSI.

3GPP and 3GPP2 have many members in common: they hold joint meetings and cooperate with each other to promote harmonization and convergence of approaches. They also work with other standards bodies to avoid duplication.

The relationship between 3GPP and IETF is formalized in the IETF RFC 3113. The intention of this is to agree that the IETF will take responsibility for changes to any RFC, such as SIP, that 3GPP identifies as requiring extension or modification to be used in 3G networks.

2.7.6   The hierarchy of standards making up 3G-324M

The 3GPP proposal for circuit-switched mobile video telephony on Release 99 3G networks is TS 26.111 V3.4.0, "codec for circuit-switched multimedia telephony service; Modifications to H.324 (Release 1999)." 3GPP2 has produced S.R0022, "Video Conferencing Services—Stage 1," which, like TS 26.111, proposes that a modified form of H.324 is used.

The 3GPP technical specification TS 26.111 is based on a variant of H.324 implemented according to Annex C of the H.324 recommendation. H.324 uses H.223 for multiplexing and H.245 for command and control of a multimedia session. H.263 and H.261 are the mandatory video codecs and G.723.1 is the mandatory audio codec that all H.324 user terminals should support.

In Annex C the modem interface of H.324 is replaced by a wireless interface, and several different multiplexer schemes are provided, each with a different level of robustness and error protection. These are referred to as mobile levels 1, 2, and 3 respectively and are more fully specified in annexes A (ML1), B (ML2), and C and D (ML3) of H.223. H.223 implemented without incorporating the features described in any of these annexes is referred to as mobile level 0 (ML0).

The main modifications to H.324M proposed in TS 26.111 V3.4.0 are that GSM-AMR should be the mandatory audio codec, support for G.723.1 becomes optional. H.263 is the mandatory video codec, H.261 and MPEG4 are optional. The multiplexer support for annexes A and B of H.223, mobile levels 1 and 2 is mandatory, mobile level 3 is not mentioned (in later releases mobile level 3 is optional). Finally the terminal must be capable of supporting a minimum bit rate of 32 kbit/s at the multiplexer to wireless interface.

The documents from 3GPP do not give detailed justifications for the modifications 3GPP has made to H.324; however, we can speculate on what they are. As a codec specifically developed for mobile applications GSM-AMR may be considered to give superior performance to G.723.1 in this application. It may also have been proposed because any handset that supports video telephony will also need to support standard speech telephony and will have a GSM-AMR codec for that purpose. Its reuse for video telephony reduces the resource requirements of the terminal. The lack of requirement to support mobile level 3 may also be driven by resource considerations—it requires a significant amount of computation, requiring a handset with a high specification or limiting the bit rate the multiplexer is capable of producing.

2.8   Chapter Summary

The basic principles behind conversational multimedia communication have been presented in this chapter, for terminals on both packet and circuit-switched networks. Compared to speech telephony more complex control is required to discover the capabilities of each terminal and to establish a multimedia session with agreed media types. It is particularly important to ensure that control messages are transmitted reliably.

Different types of information may require different treatment to ensure that they are adequately protected from errors. Transmission of information over packet networks generally uses the well known TCP/IP and UDP protocols to transport the data. On circuit-switched networks more efficient multiplexing schemes are preferable.

Through the use of an example we have looked at how control of multimedia sessions can be described in terms of signaling entities, and how ASN.1 can be used to describe the messages that flow between terminals, with PER to produce compact binary representations of these messages.

The components of a circuit-switched multimedia have been presented, identifying those areas that are within the scope of standardization.

The need for standards to allow interoperability is obvious. We concluded the chapter with a review of the standards making bodies and their different roles, and an overview of the standards making up 3G-324M.

The chapter is intended to provide background to help in understanding the specifics of 3G-324M-based video telephony, which is the subject of most of the rest of the book, and SIP-based video telephony presented in Chapter 9.

3

Media Coding

The purpose of the codecs within a multimedia terminal such as a mobile video telephony handset is to take the captured media information and convert it into a compressed digital format suitable for transmitting over a channel with restricted bandwidth, and to take compressed transmitted media information and decode it to construct an approximation of the original signal.

This chapter reviews the techniques that are used for the compression of audio and video signals. It describes the coding standards that are most relevant to mobile video telephony in sufficient detail to provide a high level understanding of how they are integrated into a 3G-324M compliant terminal and controlled at the system level.

3.1 Principles of Speech Coding

When the telephone network was first developed it was decided that an analog speech signal limited by means of filters to 4 kHz provided adequate quality to be intelligible and for the speaker to be recognized. This decision has been accepted ever since and has influenced the design and dimensioning of subsequent telecommunications systems. When digital techniques replaced analog, the speech was sampled at the Nyquist rate to give 8,000 samples per second and then encoded using 7 or 8 bits per sample to give a 56 or 64 kbit/s PCM signal. This is the reason that most modern circuit-switched telecommunications networks switch bandwidth in 64 kbit/s chunks.

For video telephony to be possible within this limitation of 64 kbit/s bandwidth the speech signal must be significantly further compressed to provide some headroom for the video signal. The audio codecs used in mobile video telephony are intended for encoding speech, and so are designed to exploit properties of the speech signal. Many techniques have been developed, but the codecs used in video telephony are based on the same general algorithmic approach, known as *code excited linear prediction* (CELP).

CELP is based on the idea that speech is a noise made by humans using their vocal chords to produce an excitation signal, which is then shaped or filtered using a combination of their nose, mouth, lips, and tongue. If the speech signal can be analyzed to extract the excitation signal and a set of filter parameters, these can be transmitted and used at the receiving end to regenerate the speech. The filter changes on a timescale of tens of milliseconds—corresponding to changes in the mouth shape and tongue and lip positions. The approach of periodically extracting the excitation signal and the filter parameters and encoding and sending them can lead to a significant reduction in the bandwidth needed to transmit speech, compared to 64 kbit/s required for PCM speech.

Figure 3.1 is a very simplified high-level block diagram of a CELP encoder and decoder. It shows that in the encoder a frame of incoming speech samples is analyzed to extract the excitation and filter parameters. These are applied to a synthesis block that is effectively a filter modeling the human vocal tract, to regenerate the speech. This is subtracted from the incoming speech samples to refine the estimate and minimize the residual signal (the output of the adder in Figure 3.1). The process is known as analysis by synthesis.

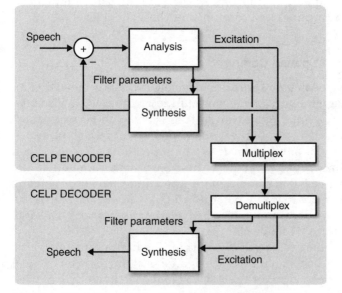

Figure 3.1 Block diagram of a CELP-based speech codec.

The excitation can be one of two types, voiced or unvoiced. Unvoiced excitation is analogous to the excitation used by humans when making sounds like "ssss," or at the beginning of "te" or "ku" and is a wide spectrum noise signal. Voiced excitation is used for "eeee" and other vowel like sounds and is a near-periodic signal with an associated pitch, which must also be encoded and sent.

The encoder may operate in a number of modes with different bit rates, so control signals may be used to signal to the decoder the mode that the encoder is using.

When the parameters, excitation type, and pitch information for the frame have been extracted they are combined into output frames by the multiplex component shown in Figure 3.1. This may also add control signals if the codec uses them, and apply any extra bits required for error protection purposes.

At the decoder the excitation and the parameters are used to resynthesize the speech.

All the various CELP-based standards use variants of this underlying approach. The size of the speech frame analyzed varies from 10 ms to 30 ms or so for the different standards, and the codec may have a number of modes of operation in which it produces different amounts of compression and therefore encoded bits per frame. There may be a facility to detect when there is no speech present in the input frame. If silence is detected no encoded bits need be sent. During this period the decoder may generate comfort noise—low level background noise to reassure the listener that the line has not simply dropped.

3.2   The G723.1 Speech Codec

The ITU-T G.723.1 recommendation "dual rate speech coder for multimedia communications transmitting at 5.3 and 6.3 kbit/s" was produced for use with H.324, although it has achieved popularity in other applications including VoIP.

As the title of the recommendation makes clear, it has two transmission rates: a high rate at 6.3 kbit/s and a low rate at 5.3 kbit/s. The codec processes speech in 30-ms frames consisting of 240 samples, and outputs either 24 or 20 octets of compressed speech, depending whether high rate or low rate is used. The rate used to encode a particular frame is signaled using a bit in the output frame, and can change from frame to frame.

The decoder reconstructs the speech signal frame by frame from the received compressed data. It is capable of performing error concealment if it gets an external indication that a frame is corrupted.

Annex A of G.723.1 describes a scheme for *voice activity detection* (VAD) and using this to provide a *discontinuous transmission* (DTX) mode where encoded frames are not sent when there is no voice activity.

During periods where no speech is transmitted, the codec sends *silence insertion descriptor* (SID) frames that are 4 octets long. The first 2 bits of the first transmitted octet of each output frame indicate whether the frame is high rate, low rate, or a SID frame.

Methods of providing progressive levels of error protection, at the cost of progressively increasing bandwidth requirements, are specified in Annex C of G.723.1.

3.3 The GSM-AMR Speech Codec

GSM 06.90 adaptive multirate speech transcoding, (known more commonly as GSM-AMR) was specified by ETSI for use in GSM-based mobile systems. The full specification is covered by additional documents: GSM 06.73 (which provides reference code in C) and GSM 06.91–GSM 06.94. These deal separately with frame loss, comfort noise, discontinuous transmission, and voice activity detection.

GSM-AMR operates at eight different bit rates: 12.2, 10.2, 7.95, 7.4, 6.7, 5.9, 5.15, and 4.75 kbit/s. When used within 3G-324M this corresponds to output frames of 31, 26, 21, 19, 18, 16, 14, and 13 octets respectively. These frames include CRC bits for error detection. The mode used to encode the frame is also signalled in each frame and can be changed from frame to frame.

GSM-AMR also supports VAD. When silence is detected two 6-octet SID frames are sent in successive 20-ms intervals (the SID_FIRST and SID_UPDATE frames) followed by 1-octet DTX frames (0xFF) for the remainder of the duration of the silence period. The SID frames contain information on the characteristics of the comfort noise that the decoder should generate during the silence period.

If a frame is lost in transmission, the decoder can substitute the lost frame with a predicted frame that it computes from previously transmitted frames.

3.4 Principles of Video Coding

A digital video picture, or frame, is made up of picture elements or pixels. Uncompressed video requires significant amounts of bandwidth. Consider a frame that is 176×144 pixels in size. If each pixel value is encoded using 8 bits this represents over 200 kbits of information. If 10 frames per second are transmitted in this uncompressed form, 2 Mbit/s of bandwidth is needed. Video needs to be compressed at much higher ratios than speech. G.723.1 represents approximately 10:1 compression compared to 64 kbit/s PCM speech. We need to compress the video signal that we have just described by a ratio of over 40:1 to get it down to around 50 kbit/s to make it possible to transmit video along with audio (using GSM-AMR or G.723.1) in a 64 kbit/s digital channel. In fact the compression ratio is higher than this because we have not taken into account that the video has luminance and chrominance (color) components.

In the video standards we are interested in, a number of techniques are used in combination to compress video. We will look at some basic terms for describing video pictures and then briefly review the techniques used.

3.4.1   Describing video

Video is made up of a series of pictures or frames that are presented in sequence relying on the persistence of human vision to create the illusion of motion. The rate at which they are presented must not fall below 5–6 frames per second or they will look like a sequence of still images: ideally the rate should be higher than that to get video that does not look jerky.

If there is a fixed bandwidth available for transmitting encoded video there is a clear trade-off between the number of bits used to code each frame and the number of frames that can be encoded and sent per second. The higher the frame rate, the fewer bits per frame and therefore the lower the quality of each frame. The optimal setting for this trade-off is dependent on the content being encoded: a fast moving sports sequence may mean that a higher frame rate with lower quality per picture is preferable to see the flow of the action, whereas a slowly panning sequence from a travel documentary may appear better with a lower frame rate and better quality per frame.

Standardized picture sizes are based on multiples and submultiples of the *common intermediate format* (CIF). A CIF sized image is an array of 288 rows of 352 pixels. A *quarter CIF* (QCIF) image has each of these figures halved, to give an array of 176×144 pixels. *Sub-QCIF* (SQCIF) is not a simple submultiple, it is 128×96 pixels. The 4CIF picture size represents a doubling of the CIF linear dimensions and 16CIF a doubling again. In 3G-324M video telephony QCIF and SQCIF are the sizes of most interest and relevance.

Each frame is divided up into subelements that are used for encoding and transmission. Figure 3.2 shows a QCIF picture. The smallest element is the pixel. A block is made up of an 8×8 array of pixels. *Macroblocks* (MBs) are 2×2 arrays of blocks (16×16 arrays of pixels). The QCIF picture is divided into nine *Groups of Blocks* (GOBs). The size of a GOB is dependent on the size of the picture: it is normally a horizontal row of MBs the length of the picture. Alternative ways of encoding the picture may replace fixed length GOBs with variable length slices.

The output of a color camera can be represented as three components in a number of ways. One of these is as *red, green, and blue* (RGB) components. An alternative representation is as a luminance signal Y, which is effectively a greyscale version of the video, and two color difference

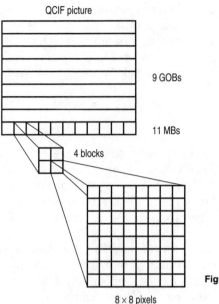

QCIF picture

9 GOBs

11 MBs

4 blocks

8 × 8 pixels

Figure 3.2   Structure of a QCIF picture.

signals CB and CR. These signals can be treated separately and processed in parallel using the same techniques for each signal. The only difference is that the human eye is much less sensitive to these signals than it is to the luminance signal, so they can be sampled at half the rate in both directions. Therefore the CB and CR picture sizes for a QCIF image are 88 × 72 pixels in size, and a macroblock is fully defined as consisting of four luminance blocks and one block from each of the color difference components.

3.4.2   DCT-based compression

Speech is a one-dimensional signal; the amplitude of the signal varies with time. A picture is a two-dimensional artifact, the value of the pixels varies with their position (x, y) in the picture. The information in a speech signal can be represented in the frequency domain by performing a Fourier transform on it. This reveals structure that is not apparent in the time domain. It can then be modified by removing some frequency components, for example by low pass filtering. This removes information and allows it to be represented by fewer bits per second.

The first technique for compressing video is analogous to this in two dimensions. The picture is made up of spatial frequencies. Figure 3.3 shows some examples of blocks of the picture containing patterns that

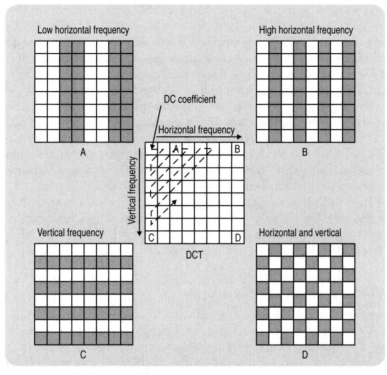

Figure 3.3   Spatial frequencies and the DCT.

represent spatial frequencies in either the horizontal (A, B) or the vertical dimension (C), or both (D). The *discrete cosine transform* (DCT) takes each 8×8 pixel block in the picture and transforms it into another 8×8 block of values representing the spatial frequency components in the block—this is illustrated by the central block in Figure 3.3. This shows in an illustrative way the areas of the DCT that would have nonzero values for each of the patterns A, B, C, and D. The top left hand coefficient of the DCT is the DC coefficient. If all the pixels in a block of the picture have the same value, this coefficient would be the only nonzero value in the DCT block. All the examples in the figure will have some DC component, representing the average value of the pixels in the block.

The DCTs of real picture blocks tend to have significant values only in the upper left-hand side of the DCT, with values in the lower right usually being small or zero. The human eye is less sensitive to this higher frequency information. If coarse quantization is performed, which is broadly equivalent to low-pass filtering, many of the DCT coefficients become zero. This results in a significant reduction in the information that must be sent to represent the block, with limited impact on perceived quality.

The results of the DCT and quantization operations can be compressed further by scanning them out in a zigzag order as indicated in the DCT block of Figure 3.3. This produces a sequence with relatively few nonzero values separated by runs of zero values. It can be encoded as a set of {runlength, value} pairs, where runlength is the number of zeros preceding a nonzero value, and codes made up of variable numbers of bits can be assigned to each pair, where the number of bits required is inversely proportional to the probability of occurrence of a particular {runlength, value} pair. This is a lossless or reversible coding technique known as *variable length coding* (VLC).

Video frames that are encoded without using any information from other frames are known as Intraframes or I-frames. These are encoded using just the DCT, quantization and variable length coding, and without reference to other frames in the sequence.

3.4.3 Motion compensation

There is usually a high degree of similarity between successive frames so this can be exploited. If we consider a human head and shoulders scene with a stationary camera, the background may be the same from frame to frame. We can subtract the current frame from the previous frame or, to avoid accumulating errors, from the decoded representation of the previous frame before coding it using the DCT-based method we have just described. This reduces the information that must be coded.

An extension of this approach is to look for blocks in the decoded representation of the previous frame that best match the block that is currently being encoded, and perform the subtraction using that block. Using this technique means that information on the offset that has to be applied must be sent to the receiver. These offsets are known as motion vectors. The process of computing them is called motion estimation, and their use in the decoding process to reconstruct the picture is called motion compensation.

Because the search and compare operation is computationally intensive, the search is usually restricted to an area in the vicinity of the block being coded. This also limits the maximum size of motion vectors that can be generated. Use of interpolation between pixels allows noninteger motion vectors to be computed.

The use of motion compensation prior to computing the DCT produces a significant further reduction in the bits needed to encode a frame. Frames encoded in this way are known as Interframes, predicted frames, or P-frames.

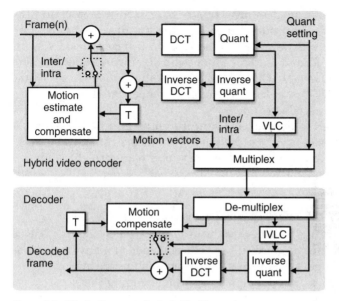

Figure 3.4   Block diagram of a hybrid video codec.

3.4.4   The hybrid video codec

Codecs that use this range of different techniques to encode video are known as hybrid codecs. A high level schematic diagram of a generalized hybrid video encoder and decoder is shown in Figure 3.4. There are control signals defining whether the frame is to be encoded in inter or intra mode and indicating the quantization settings to be used. The quantization settings may vary within the frame.

If the coding of a frame is performed in intra mode the incoming frame goes through the DCT, quantizer, and variable length coding processes and the resulting compressed information is multiplexed for transmission to the decoder. Information on the quantization used and an indication that the frame has been encoded in intra mode also needs to be transmitted. The encoded frame is also decoded within the encoder to provide a reference picture for use in encoding P-frames.

To encode a frame in inter mode, the incoming frame is compared to the reference picture of the previous frame (the block marked T in Figure 3.4 represents a delay of one frame). Motion vectors are computed, and a motion compensated version of the reference picture is subtracted from the incoming frame prior to further encoding. The remainder of the process is identical to that of an I-frame, except that for a P-frame the motion vectors must also be transmitted along with the frame type and the quantizer settings.

At the decoder received information is extracted to determine the frame type, the VLC encoded video and the quantizer settings and obtain any motion vectors. The *inverse VLC* (IVLC) converts the information back into a format suitable for decoding by inverse quantization, and computing the inverse DCT. If the frame being decoded is a P-frame the output of the DCT is added to the motion compensated reference picture to obtain the decoded frame.

The source video to be encoded is captured at up to 30 frames per second. The video rate produced by the codec can be controlled by selectively dropping input frames, and by varying the quantization applied to the output of the DCT to vary the number of bits per frame.

3.4.5   Algorithmic variants of hybrid coding

The hybrid codec described above is the basis of the video standards used in 3G-324M. In these standards it has been extended in many ways, some of which are fundamental features of a particular standard and others are options that can be selected dynamically when running the codec.

Some examples of algorithmic variants include:

- *Bidirectionally predicted frames* (B-frames) may be used to increase frame rates by interpolating between two P-frames or between an I-frame and a P-frame.

- The motion vector search area may vary, more sophisticated forms of motion estimation may be used with multiple motion vectors per macroblock and different resolutions of subpixel motion vector accuracy may be provided.

- Intra coding may be performed on selected macroblocks within a frame that is otherwise predicted. An I-frame contains many more bits than a P-frame, so refreshing a few macroblocks at a time over a number of frames will reduce fluctuations in bandwidth requirements.

- Different schemes may be used for quantization.

- Instead of partitioning each picture into GOBs, a more flexible slice-based approach can be used, where the beginning and end of the slices do not need to correspond to the edge of the frame.

- Instead of VLC other algorithms may be used to perform the lossless coding of the quantized DCT coefficients.

3.4.6   Bit-stream formats for video

The multiplex process in Figure 3.4 combines the encoded information and encoder settings to form a defined bit-stream structure. A possible bit-stream structure is illustrated in Figure 3.5. This shows a multilayered structure, with a picture layer, a GOB layer, and a macroblock layer.

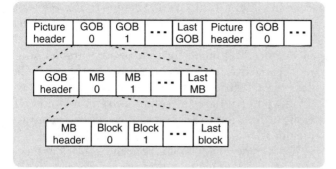

Figure 3.5 Structure of a video bit stream.

The picture header will typically provide a code to allow the decoder to synchronize with the bit stream by detecting the start of the picture, and will provide information on such things as the timing of the frame relative to the previous frame, what type of frame it is (e.g., I-frame or P-frame), what size of frame it is, what algorithmic options have been used in encoding it, and quantizer information.

Each GOB (or slice, if used instead of GOBs) will also have a header, which may include a start of GOB code for synchronization purposes, quantizer information, the number of the GOB, and therefore its position in the picture.

At the macroblock layer the header information may include whether it contains any data (it may be identical to the macroblock in the same position in the previous frame) and if so whether it is encoded in intra mode or not and any changes to quantizer settings. If it is a predicted macroblock, motion vector values will be included. After the header, any encoded information for the four luminance blocks and two chrominance blocks is provided.

In the bit-stream structure described it can be seen that there are opportunities to provide updated quantizer settings at many points.

3.5 The H.263 Recommendation

The ITU-T H.263 recommendation "video coding for low bit-rate communication" is based on the principles we have described. It supports CIF, QCIF, SQCIF, 4CIF, and 16CIF picture sizes. Motion vectors are computed per macroblock with a half-pixel accuracy and are restricted to a 16×16 pixel range centered on each macroblock. Every macroblock must be encoded and sent in intra mode at least once every 132 frames.

A codec that conforms to the main body of the recommendation is known as baseline H.263. The recommendation has a large number of annexes specifying different algorithmic enhancements, not all of which are compatible with one another. 3GPP has specified baseline H.263 as the mandatory codec for 3G-324M. It also recommends that annexes I, J, K, and T of the recommendation should be supported. These annexes are referred to as H.263 Profile 3.

Annex I, advanced intra-coding mode, introduces the prediction of macroblocks within an I-frame from neighboring macroblocks within the frame. It also provides a modification to the quantization and because the likelihood of occurrence of {runlength, value} pairs in a quantized DCT block for an I-frame is different to a P-frame it specifies a separate VLC for intra-coded frames.

H.263 coded images often show visible artifacts at macroblock or block boundaries leading to the appearance of a grid superimposed on the image. Annex J specifies the inclusion of a deblocking filter, which reduces these artifacts and improves coding efficiency.

Annex K provides a slice rather than GOB-based approach to partitioning and encoding pictures.

Annex T describes a modified quantizer that provides different quantization for luminance and chrominance values and other enhancements that lead to greater coding efficiency.

Other Annexes of interest (because as will be seen, they are referred to in H.245 messages relating to video) are annexes D, E, F, and G. Annex D allows unrestricted motion vectors—a way of computing motion vectors for MBs on the edge of the picture that gives superior performance. Annex E specifies the use of arithmetic coding instead of VLC. Because use of arithmetic coding is subject to the payment of patent related royalties it is rarely used. Annex F specifies an advanced prediction mode, which allows four motion vectors per MB rather than one, to allow predicted MBs to be composed from superimposed MBs from the reference picture. Annex G describes a PB-frames mode that introduces B-frames to increase frame rate. These are encoded and sent with P-frames as a single unit.

The bit stream produced by an H.263 encoder is substantially similar to that shown in Figure 3.5: the code used to indicate the start of a picture is known as the *picture start code* (PSC).

H.261 is an optional codec for 3G-324M. This can be viewed as a less efficient predecessor to H.263. It was probably included with the intention of facilitating interoperation between 3G mobile video services and legacy ISDN-based H.320 services (where H.261 is the mandatory codec). It is unlikely that many 3G-324M compliant terminals will support the H.261 option.

3.6   The MPEG4 Video Codec

The ISO/IEC 14496-2 standard "Information Technology—Coding of Audio-Visual Objects—Part 2: Visual" is better known as MPEG 4 or MPEG4 visual. MPEG4 is an optional video codec for 3G-324M terminals.

The principle behind MPEG4 is that an image can be decomposed into a set of objects, and different tools can be used to independently encode each object. These can include model-based coding requiring sophisticated techniques such as recognition of objects within a scene and modeling them using wire frames. In the decoder these objects are reconstructed and superimposed to create the image. Profiles and levels are defined that specify which set of tools is to be used. We need not be concerned about the more sophisticated techniques available in MPEG4 because Simple Profile@Level 0 is used in 3G-324M.

MPEG4 Simple Profile@Level 0 is broadly similar to H.263. It supports one video object that is rectangular in shape and has a maximum size of 176 × 144 pixels, corresponding to QCIF, with a maximum frame rate of 15 frames per second. From now on when we use the term MPEG4, we will be referring to this profile and level.

Baseline H.263 is a subset of this profile and level, and is known as MPEG4 short header. From this it is clear that the structure of the bit stream is very similar to that used in H.263. There is some change in the terms used to describe it. MPEG4 uses the concept of *video object plane* (VOP). An I-VOP is analogous to an I-frame, a P-VOP to a P-frame, and a B-VOP to a B-frame.

Some features of MPEG4 that are enhancements to H.263 baseline are *data partitioning* (DP), *reversible VLC* (RVLC), resynchronization markers, and header extension codes.

In H.263 the PSC and the GOB start codes can be used by the decoder to resynchronize with the bit stream when transmission errors occur. Because video compression is intrinsically variable, these codes occur at irregular intervals in time. The resynchronization markers in MPEG4, known as video packet headers, can be inserted between any two macroblocks. This allows the encoder to control the maximum amount of encoded information that can be sent before a resynchronization marker occurs. A similar approach can be taken in H.263 by using the slice approach of annex K.

Data partitioning involves separating the components of the macroblocks in an MPEG-4 video packet. Without data partitioning, each fully coded MB is sent in order. With data partitioning in each slice or video packet the headers of all MBs are sent first, then the motion vectors and finally the quantized DCT coefficients.

The quantized DCT coefficients in a video packet can be encoded using RVLC. Each codeword used in VLC may be a different number of bits long. If an error occurs in a sequence of VLCs it can become impossible to work out where one codeword ends and the next begins after the point where the error occurs. RVLC provides a set of codes that allow the boundaries between individual code words in a sequence to be detected by examining the sequence from either end. If an error occurs when RVLC is being used the whole sequence of code words does not need to be discarded. Using the markers the block can be examined from the other end to extract the code words. This gives greater resilience to errors.

Header extension codes are used to repeat the information contained in the VOP header elsewhere within the VOP, allowing the information to be decoded even if the VOP (frame) header is lost. This means that if this does occur, the whole VOP does not have to be discarded.

3.7 Chapter Summary

In this chapter we have outlined the principles behind the audio codecs and video codecs used in 3G-324M. The audio codecs are all based on the underlying CELP approach. The video codecs are known as hybrid codecs, and are based on underlying toolsets that employ similar principles, to the extent that baseline H.263 is a subset of MPEG4. Specific features of GSM-AMR, G.723.1, H.263, and MPEG4 have been discussed.

The information presented in the chapter is intended to provide background information that, in later chapters, will assist in understanding how codecs based on these recommendations and standards are used and controlled within a 3G-324M terminal.

Chapter 4

Video Telephony over Mobile Switched Networks

We have said that for 3G mobile networks conforming to Release 99, conversational video telephony services are based on 3G-324M, a derivative of the existing ITU-T recommendation H.324 for low bit-rate media communication, which was initially intended for video telephony using modem-based communication over the GSTN. In this chapter we will explore the H.324 recommendation including Annex C that is referred to as H.324M, and the derived specification of 3G-324M.

The main recommendations that 3G-324 refers out to and depends on will be introduced, including H.223 for multiplexing and H.245 for command and control. We will look at the steps that a 3G-324M terminal goes through in establishing a session and the techniques that are used to deliver control messages reliably, and conclude by briefly looking at the establishment of a bearer for a 3G-324M call.

4.1 The H.324 Recommendation

The main body of recommendation H.324 "terminal for low bit-rate multimedia communication" covers the components shown in Figure 4.1. These components map straightforwardly onto the components that are within the scope of multimedia standards in Figure 2.15.

One of the key concepts of H.324 (which is also used in H.323) is that of logical channels. *Logical channel number zero* (LCN0) is dedicated to control information. It is implicitly opened as soon as a connection has been established, using an associated multiplexer frame structure that allows LCN0 information to be transmitted and received. Using LCN0 the terminal can immediately start sending and receiving control

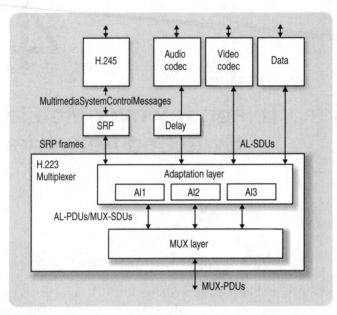

Figure 4.1   Components of an H.324 terminal.

information following the syntax and semantics defined in ITU-T recommendation H.245, allowing it to declare its capabilities and discover the capabilities of the remote terminal. Both terminals then use this information to open additional logical channels that are consistent with the declared capabilities.

Another key feature is the approach taken to multiplexer frame structures—the way that octets in each frame are allocated to control, media, and data information. Each terminal can have up to 16 multiplexer frame structures. One of these is predefined for use with LCN0 at the start of the session. The structure of the other multiplexer frames that the terminal will use is signaled to the receiving terminal at the start of the session, in the form of *multiplex table entries* (MTEs). There is no requirement for a terminal to send or use all 15 MTEs, and communicating terminals are not restricted to using a common set of MTEs. The MTE that is used for a particular multiplexer frame is indicated via a *multiplex code* (MC) value contained in the header of that frame.

H.223 provides the multiplexing function for H.324. It has a two-layer structure: the upper *adaptation layer* (AL) provides three ways of adapting media, data, and control information (known as AL1, AL2, and AL3), which provide increasing levels of error protection. The lower multiplexer layer takes information from different sources passed to it by the adaptation layer and combines it to form multiplexer frames. It also performs the reverse operation on incoming frames.

The *simple retransmission protocol* (SRP) layer shown in Figure 4.1 between the H.245 control process and the multiplexer is designed to provide reliable, acknowledged transmission of control information.

The mandatory video codecs for H.324 are H.261 and H.263. The allowed resolutions are CIF, 4CIF, 16CIF, QCIF and SQCIF. For modem-based communication, or even 64 kbit/s digital-switched circuits, use of resolutions above QCIF will lead to either very low frame rate or very poor picture quality so in practice these are unlikely to be used for conversational multimedia, though they may be useful for more specialized applications such as slow scan surveillance. The mandatory audio codec is G.723.1, supporting both 6.3 kbit/s and 5.3 kbit/s rates and transmission of silence frames.

There is an optional delay that can be inserted in the receive audio path to equalize delay in the video and audio paths and ensure that lip synchronization is maintained. In general it is expected that delay in the video path will be greater than for audio, because the video codec has greater algorithmic delay and computational complexity than the audio codec. The amount of delay which is inserted in the audio receive path can be set by a H.245 message, H223SkewIndication, from the transmitting terminal.

H.324 also covers procedures for interfacing to a synchronous modem and call initiation using the modem, but these are not discussed here because they are not used in the derived 3G-324M. Readers interested in this aspect should refer to the recommendation itself.

4.2   H.324M

Annex C of H.324 was introduced as a result of studies on how H.324 could be adapted for use over wireless and mobile networks, initially focusing on such technologies as *digital enhanced cordless telecommunications* (DECT). H.324 with Annex C has become known as H.324M—Figure 4.2 is a block diagram of the components of an H.324M terminal. Annex C removes the modem requirements for H.324 and assumes that a transparent digital channel is available. Annex C provides three possible additional modes for more robustly multiplexing information: these are known as mobile levels 1, 2, and 3. Mobile level 0 is the basic mode of operation of the multiplexer in H.324 without Annex C. Annex C of H.324 defines a mobile level detection mechanism to allow terminals to synchronize at the start of a session and establish which of the possible mobile levels is to be used to send multiplexed information. A mechanism for changing mobile level during a session is provided.

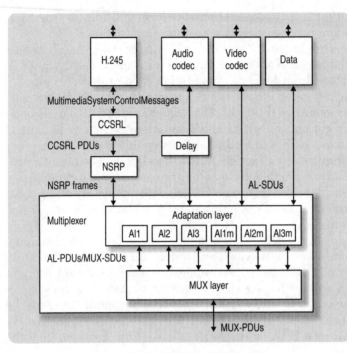

Figure 4.2 Components of an H.324M terminal.

Annex C of H.324M also provides enhanced procedures for more reliable delivery of H.245 control messages by specifying the use of *numbered SRP* (NSRP) instead of SRP. A layer known as the *control segmentation and reassembly layer* (CCSRL) is introduced between the H.245 signaling entities and the NSRP layer. The purpose of CCSRL is to segment larger control messages, to reduce the susceptibility of control messages to error.

The only change to the mandatory codecs is that support for Annex C of G.723.1, "scalable channel coding scheme for wireless applications," is recommended.

4.3   The 3G-324M Recommendation of 3GPP

3GPP has taken H.324M as a starting point and modified it to create 3G-324M. For Release 99 the principal differences between 3G-324M and H.324M are in the codecs supported and in the limiting of mandatory support by the multiplexer to mobile levels 1 and 2. Mobile level 3 is optional: use of this level may make too high a computational demand on the handset.

4.5   Introduction to the H.223 Multiplexer

The need to multiplex information for multimedia services over switched-circuit networks was discussed in Chapter 2. The multiplexer used in H.324, H.324M, and 3G-324M is specified in ITU-T recommendation H.223 "multiplexing protocol for low bit-rate multimedia communication."

The multiplexer takes video and audio, control messages, and any other data in logical channels; and forms multiplex frames for transmission in a format that is understood by the receiving terminal. It also demultiplexes information sent to it by the remote terminal to extract its contents for processing, allowing two-way conversational multimedia information transfer. The different types of information that are multiplexed may have different properties, this means they may need to be handled in different ways, with error checking, transmission guarantees, and facilities for retransmission, if any, that are appropriate to each information type.

We have seen in Chapter 2 that a two-layer structure, consisting of an upper adaptation layer and a lower *multiplex* (MUX) layer enables a multiplexer to handle different types of information in different ways. The H.223 multiplexer adopts this approach, as shown in Figures. 4.1 and 4.2.

The adaptation layer provides a way of handling different information types appropriately. It takes information from various sources (H.245 Control, audio codec(s), video codec(s), and any application data sources) and adds supplementary information to form *adaptation layer protocol data units* (AL-PDUs). These AL-PDUs are then passed to the MUX layer.

The MUX layer assembles AL-PDUs into chunks of data called *MUX protocol data units* (MUX-PDUs), adding header information to indicate how the contents of the PDU have been assembled, and providing some error detection capabilities on the header information. These MUX-PDUs are then sent via the network to the remote terminal.

In the reverse direction MUX-PDUs are extracted from the bit stream received from the remote terminal and demultiplexed by the MUX Layer to reassemble the AL-PDUs. These are then passed up to the adaptation layer that in its turn extracts the source information and passes it up to the higher-level component that will process it.

The terms used to describe the units of information flowing into and out of the multiplexer, and between the adaptation layer and the MUX layer can be confusing to someone reading the H.223 recommendation for the first time. They are as follows:

AL-SDU (*adaptation layer service data unit*) is the term used to refer to the information that is exchanged between the adaptation layer and

The mandatory audio codec for 3G-324M is GSM-AMR and suppo⟩ for G.723.1 becomes optional. H.263 remains a mandatory video code⟨ H.261 becomes optional and MPEG4 Simple Profile Level 0 is added ⟨ the list of optional codecs.

4.4 Logical Channels

It was stated earlier that logical channels are a key concept in H.32 When a multimedia session is started the predefined LCN0 a⟩ MTE0 are used to transmit and receive H.245 Multimed SystemControlMessages. These messages include requests to open (a later to perhaps close) additional logical channels for video, audio, a data. The MTEs are defined in terms of logical channels, for exam MTE1 may be defined as referencing a multiplexer frame structure t⟩ contains 20 octets of LCN1 followed by LCN2 until the closing flag t⟩ delimits the end of the frame.

An important property associated with each logical channel is whet or not it can be segmented. This refers to whether or not the infor⟩ tion that the logical channel is carrying can be broken up into f⟩ ments for multiplexing and transmission. Whether a logical channel be segmented or not is specified as part of the process of establishin⟩ opening a logical channel.

The process that supplies the source information for the logical c⟨ nel may provide this information with some intrinsic structure. output of an audio codec may be periodic small frames representing 2 of encoded speech. A video codec may provide encoded video as ⟩ larger chunks of information but less frequently, perhaps every 10 If the speech frames have to wait for a complete video packet to be before a multiplex frame can be formed and sent, this will lea⟨ 100-ms delay to the audio. Allowing video to use a segmentable ⟩ channel reduces problems with audio delay and provides a less load to the multiplexer.

A further reason for using segmentable logical channels is t⟩ natural size of the chunks of data being output from a video c⟨ other process using the multiplexer may be larger than the ma⟩ size of a multiplexer frame. This is especially relevant to mobil⟨ media terminals where multiplexer frame sizes must be kept ⟨ reduce the likelihood of errors in the channel corrupting the fr⟩ general segmentable logical channels are used when the info⟩ flows at variable rates, as is the case for control, video, a⟩ Nonsegmentable logical channels are used where the informa⟩ is regular and fairly low bit rate, such as speech. A segmentab⟩ channel must also be used where its input does not have intrin⟩ ture but is effectively a stream of bits or octets of indefinite d⟩

the codecs, such as H.245 control or other processes that are making use of the services of the multiplexer.

AL-PDU (*adaptation layer protocol data unit*) is a term used to refer to the information that is exchanged between the adaptation layer and the MUX layer, usually when discussing the operation of the multiplexer from the perspective of the adaptation layer.

MUX-SDU (*MUX service data unit*) is a term used to refer to the information that is exchanged between the adaptation layer and the MUX layer, usually when discussing the operation of the multiplexer from the perspective of the MUX layer. MUX-SDU and AL-PDU are therefore different terms to describe the same thing.

MUX-PDU (*MUX protocol data unit*) is a term used to refer to the information that is exchanged between local and remote multiplexers. The MUX-SDUs for a segmentable logical channel can be broken up and distributed across several MUX-PDUs.

The adaptation layer has several options for processing AL-SDUs to form AL-PDUs/MUX-SDUs for presentation to the MUX layer. AL1, AL2, and AL3 (and AL1M, AL2M, and AL3M if supported) provide increasing levels of protection for the AL-SDU data they contain. Different adaptation layer options can be applied to different logical channels.

The MUX layer has a number of ways in which it can frame data to form MUX-PDUs, which provide greater levels of robustness in the face of increasingly error prone network connectivity. These are referred to as mobile levels 0, 1, 2, and 3 (ML0, ML1, ML2, and ML3). At any time the MUX layer operates at one mobile level. It cannot dynamically mix MUX-PDUs formed according to the rules of different mobile levels; however, it is possible for the terminal to switch mobile levels during its operation by negotiation. 3GPP specifies that mobile video terminals must support ML1 and ML2. Support for ML0 is a mandatory requirement of ITU-T H.324 Annex C and therefore by implication of 3G-324M.

4.5.1 The adaptation layers

The adaptation layer types are intended for use with different types of information and provide varying degrees of error protection. AL1 is intended for control or data information. AL1 is the simplest of the adaptation layers. It does not provide any error protection capabilities or any procedures for retransmission, so it is used where these have been provided by the layer supplying the AL-SDUs; AL1 does not add any additional octets to the incoming data, so it is effectively transparent.

AL2 takes incoming AL-SDUs and appends a 1-octet CRC to provide a way of detecting errors in received AL-PDUs. AL2 also optionally provides a 1-octet header containing a sequence number that is incremented with every AL-PDU formed. This enables a receiving terminal to detect missing AL-PDUs, or PDUs delivered in the wrong order (this is unlikely in a circuit switched network, but not impossible given that the communications in the RAN are ATM-based). According to H.223 AL2 is intended to be used for audio.

AL3 is intended for the transport of video. It adds an optional control field header to the AL-SDU and appends a 2-octet CRC computed over the header and payload. The control field may be 1 or 2 octets and allows an optional retransmission procedure to be implemented. It comprises a 7- or 15-bit (in the case of 2-octet control fields) sequence number and a single-bit *payload type* (PT) field. The PT field indicates whether the AL-PDU carries an AL-SDU or a message related to the retransmission procedure.

Adaptation layers AL1M, AL2M, and AL3M are specified for use with ML3. They provide error correction capabilities but are considerably more computationally complex than AL1, AL2, and AL3 and would be very demanding on any handset that made use of them. ML3 is not described further because ML3 is not recommended for 3G-324M for Release 99.

4.5.2   The MUX layer

The MUX layer takes AL-PDUs (MUX-SDUs) and combines them to form MUX-PDUs. The MUX-PDUs are separated by opening and closing flags that are not considered to be part of the MUX-PDU. When it has no complete MUX-PDUs ready to send the multiplexer must maintain the synchronous bit stream. To do this it sends out a stuffing sequence. The multiplexer layer forms and handles MUX-PDUs differently for the different possible mobile levels. The differences lie in the flags used to indicate MUX-PDU boundaries and the way they are detected, the stuffing sequence used, and the header information in the MUX-PDU.

4.5.3   Mobile levels

ML0 uses a single-octet flag with the value 0x7E (binary 01111110). There is an opening flag before the MUX-PDU and an identical closing flag after, which may also act as the opening flag for the MUX-PDU that follows. When the multiplexer has no completed MUX-PDUs to send, it sends repeated 0x7E flags as stuffing.

ML0 has a single-octet header divided into three fields. The most significant 3 bits of this are a HEC field containing a CRC for the MC field contained in the next four most significant bits positions. The MC field specifies the MTE the MUX-PDU is using. The LSB is a *packet marker* (PM), which is used to indicate whether the previous MUX-PDU contained the last MUX-SDU for a segmentable logical channel.

In ML1 a 2-octet opening and closing flag, 0xE14D, is mandatory. An optional double flag can be used, consisting of a repeat of the mandatory flag. The header structure for the MUX-PDU is the same as for ML0. Stuffing is performed by repeating the flag, so if the double flag option is used any stuffing will be an even number of 0xE14D values.

ML2 uses the same opening and closing flag as ML1, but does not support the double flag option. The header field is 3 octets or optionally 4 octets long. The 3-octet version comprises an 8-bit *multiplex payload length* (MPL) field, a 4-bit MC field with the same meaning as for ML0 and ML1 and a 12-bit parity check field for the header instead of the 3-bit CRC used in ML0 and ML1, which provides error detection and correction.

Unlike ML0 and ML1, in the mandatory header structure for ML2 there is no PM flag to indicate that the previous MUX-PDU contained the last segment of a segmentable logical channel. This function is implemented by 1s complementing the closing flag.

The optional 4-octet header for ML2 improves robustness by adding an additional octet to the header, which has the same format as the ML0 and ML1 headers but contains values for the previously transmitted MUX-PDU.

Instead of using repeated opening flags as stuffing when there is no completed MUX-PDU available for sending, ML2 sends empty MUX-PDUs with the header field set to zero separated by flags, giving 0xE1 4D 00 00 00 as the stuffing sequence.

4.5.4 Initializing multiplexers at the start of a session

When a call is made between two 3G-324M terminals there is a possibility that they may start operating at different mobile levels. The process for establishing communication between the multiplexers at the highest common level they can operate begins with each terminal indicating the highest mobile level that it is capable of operating at. It does this by repeatedly transmitting the stuffing sequence for that level. At the same time the terminal inspects the stuffing sequence it is receiving. If the sequence it is receiving represents a lower level than it is transmitting, it changes its transmitted sequence to the detected lower level.

When both terminals are sending stuffing flags for the same level the mobile level has been established. The multiplexers can then start to look for valid MUX-PDUs in the incoming stream (using LCN0 and MTE0).

4.6   Introduction to H.245 for Call Control

The control messages for H.324, H.324M, and 3G-324M are specified in ITU-T recommendation H.245 "control protocol for multimedia communication." The H.245 document is over 300 pages long, and provides the syntax and semantics of messages required to initiate and control multimedia sessions for a number of other related standards. It includes messages relating to encryption, multipoint calling, and conferencing and support for a wide range of codecs and data protocols. A subset of the messages available in H.245 is required for 3G-324M video telephony services and terminals. H.245 message syntax is specified using ASN.1, and the messages are encoded to an efficient binary representation using PER.

H.245 uses the concept of signaling entities as introduced in Chapter 2, which are associated with particular procedures or stages in the establishment of a multimedia session. When a call is set up between two terminals the SEs in one terminal communicate with peer SEs in the other. Where the procedure is independent in each direction there will be independent incoming and outgoing SEs associated with each direction, as shown in Figure 4.3. The operation of each SE is described in H.245 using SDL, which describes the actions taken by the SE and any change of state that results from primitives exchanged between the SE and the SE user, messages exchanged between peer SEs and events raised internally to the SE, such as the expiry of timers.

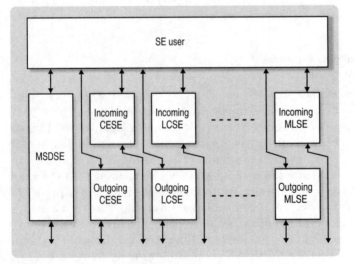

Figure 4.3   H.245 signaling entities.

The messages used between SEs in H.245 are classified as belonging to one of four types: request, response, command, and indication. Request messages require the peer SE to take action and send response messages. Response messages are of two sorts: acknowledgement messages indicating that the peer SE has accepted the request and reject messages indicating that the peer SE cannot fulfill the request. Command messages require the peer SE to take action but do not require a response. Indication messages are informational and do not require a response.

Procedures are also specified for performing various functions once the call is established and for closing the session down. All the procedures and associated SEs will be reviewed in detail in Chapter 6—for now we will look at the key procedures required to establish a 3G-324M session, which are capability exchange, master slave determination, opening logical channels, and multiplex table entry exchange.

4.6.1 Capability exchange

The capability exchange procedure is the first procedure that takes place after the multiplexers have synchronized and begun to communicate at a particular mobile level. Capability exchange uses the outgoing and incoming *capability exchange signaling entities* (CESEs). Terminals can send their capabilities to one another entirely independently in each direction. The `TerminalCapabilitySet` request message allows each terminal to signal to the other the capabilities it supports through the data provided within the structure of the request. This includes its multiplexer capabilities, supported codecs, and parameters associated with the codecs. The capability set also specifies any limitations it may have on encoded formats that it can simultaneously receive, and if there is any interdependence between the formats it can simultaneously transmit and receive.

4.6.2 Master/slave determination

In some circumstances conflict can arise when two terminals send incompatible requests. A way of resolving any conflicts is to establish that one terminal is master and the other terminal is slave. The master can then take responsibility for resolving the conflict, by rejecting the request from the slave. Master/slave determination uses the *master slave determination signaling entity* (MSDSE) and is initiated when either terminal sends the `MasterSlaveDetermination` request message. It must take place before any procedure that requires knowledge of this status becomes active: for H.324/3G-324M it happens together with or immediately after capability exchange. Master slave determination cannot be performed independently in each signaling direction. Therefore, unlike most other H.245 SEs that have independent outgoing and incoming entities, there is only a single MSDSE in a terminal.

4.6.3 Opening logical channels

Once capabilities have been exchanged and master/slave determination has completed, there is some flexibility in what happens next. In order for media to start to flow between the terminals, logical channels need to be opened. A separate instance of the outgoing and incoming *logical channel signaling entity* (LCSE) is used for each logical channel established. Logical channels are opened by the terminal that wishes to send the media sending the OpenLogicalChannel request, which specifies the logical channel number, type of channel to be opened, the adaptation layer to be used and whether or not the logical channel is segmentable.

4.6.4 Multiplex table entry exchange

Before media information can flow, the remote terminal must know how the transmitting terminal intends to format the payload of the MUX-PDUs it will use to send the data; without this information it is unable to demultiplex them. Multiplex table entry exchange uses the outgoing and incoming *multiplex table signaling entity* (MTSE) and is initiated by the MultiplexEntrySend request. This message contains a description of each of up to 15 multiplex formats and an associated MC number that is used to reference each of them.

The multiplex table entry exchange procedure may take place before or after logical channels are opened, but no media information can be sent until both procedures have completed successfully.

4.7 Reliable Delivery of H.245 Messages

For an H.324 terminal the SRP process is used to ensure reliable, acknowledged delivery of H.245 MultimediaSystemControlPDUs to the peer terminal. For 3G-324M and H.324M terminals there are two processes, between the H.245 signaling entities and the multiplexer, CCSRL, and NSRP, designed to improve the reliability of H.245 message delivery in error prone environments.

4.7.1 CCSRL

The CCSRL process is designed to improve the performance of 3G-324M mobile video terminals in conditions where the likelihood of error is high. In these circumstances larger H.245 messages may have difficulty in being reliably transmitted and received, despite the use of NSRP. CCSRL provides a mechanism for segmenting H.245 messages into message fragments.

In CCSRL each H.245 PER encoded MultimediaSystemControlPDU above a certain size (the maximum CCSRL PDU size minus 1 octet) is split into multiple CCSRL-PDUs. As shown in Figure 4.4, each CCSRL-PDU

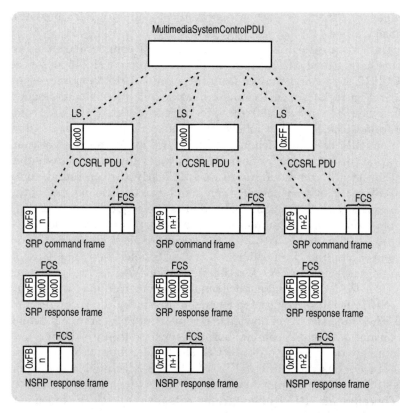

Figure 4.4   Structure of CCSRL PDUs and SRP/NSRP frames.

comprises a single octet *last segment* (LS) field followed by the CCSRL-PDU payload, which is at least 1 octet in size. The LS field is set to 0xFF if the CCSRL-PDU is the last segment of the MultimediaSystemControlPDU, otherwise it is set to 0x00.

CCSRL-PDUs are passed to the NSRP layer, where they are formed into SRP command frames. The maximum size of a CCSRL-PDU is not specified in the ITU-T H.324 recommendation or in any 3GPP documents, though clearly for the CCSRL layer to add value it should be a submultiple of 256 octets, for example 64 octets, because MultimediaSystemControlPDUs are limited to a maximum of 256 octets.

4.7.2 SRP and NSRP

The SRP is a way of providing guaranteed delivery of H.245 messages from one terminal to another. The protocol is very similar to the control acknowledgement protocol described in Chapter 2. SRP messages or "frames" are categorized as command frames and response frames.

The transmitting terminal forms and sends SRP command frames. An SRP command frame may contain one or more complete H.245 messages or if CCSRL is in use it may contain fragments of H.245 messages, as shown in Figure 4.4. An SRP command frame includes a 1-octet header, which always has the value 0xF9 to identify the start of the SRP command frame. The octet that follows this is a sequence number to allow individual SRP command frames to be identified. At the start of communication this sequence number can have any value: it is incremented every time an SRP command frame is sent. Clearly it wraps round, going from 0xFF to 0x00 every 256 SRP command frames. The payload of the frame, consisting of a CCSRL PDU, follows the sequence number. The SRP command frame is completed with a 2-octet *frame check sequence* (FCS) field, which holds a 16-bit CRC derived from the information in the frame excluding the header and the FCS field itself. The CRC is computed according to ITU-T recommendation V.42 "error-correcting procedures for DCEs using asynchronous-to-synchronous conversion."

The SRP response frames are formed and sent by the receiving terminal. They are used to acknowledge receipt of an SRP command frame. Their format is an empty command frame, consisting only of 3 octets corresponding to the header and the FCS. The header for an SRP response frame always has the value 0xFB. The FCS is also always the same (0x0000) for every SRP response frame, because the payload length is zero. Therefore an SRP response frame will always be the hexadecimal sequence FB 00 00 (see Figure 4.4).

When a terminal first sends an SRP command frame it starts (sets) a timer T401, initializes a counter to zero and waits for a response. The remote terminal is required to transmit a response within 500 ms of receiving an SRP command frame. No further SRP command frames can be sent until either a valid SRP response is received or the timer expires.

If a valid SRP response frame is received, the next SRP command frame (with an incremented sequence number) can be sent. If a valid SRP is not received before the timer expires the transmitter resends the SRP command frame, restarts the timer, and increments the counter. If the process of transmitting an SRP command frame repeats N100 times without success (a valid SRP response frame being received), the SRP process gives up and informs a higher level control process in the terminal, which takes action such as abandoning the session or call. N100 is an integer value that is determined by the implementer, though the ITU-T H.324 recommendation specifies that it should be set to a value of at least 5. Likewise, the time at which T401 expires is also not specified by the standard. It should be set to be just longer than the expected round-trip time between an SRP command being sent and an

SRP response being received. If it is less than this, unnecessary time-outs will occur often. If it is much greater, problems such as unacceptably long call setup times may arise in error prone conditions.

The SRP process as described above has a weakness. If the remote terminal is producing delayed SRP response frames, the timer will expire and the transmitter will resend the SRP command frame. An SRP response to the previously transmitted command frame may be received just after it has sent this. The transmitter then sends the next SRP command frame and mistakes the SRP response frame it receives shortly afterwards as being a response to its latest command, when in fact it is the response to the earlier retransmitted SRP command. The process is then out of step and has become unreliable. NSRP overcomes this potential problem by including the sequence number of the SRP command frame being responded to in the response frame. An NSRP response frame comprises a 1-octet header, a 1-octet sequence number that is identical to the sequence number of the SRP command frame being responded to, and a 2-octet FCS field as illustrated in Figure 4.4.

It is mandatory for 3G-324M terminals to support NSRP.

4.8 End-to-End Call Setup for 3G-324M

The first step in making a 3G-324M video telephony call is to establish the bearer. The bearer must be a transparent synchronous full duplex channel with a guaranteed 64 kbit/s of bandwidth over both the air interface and the fixed network.

The called handset, which will in general support both video and speech calls, needs to be aware that an incoming call alert is for a video or a speech call because there is no mechanism for automatic fallback to speech in the 3GPP Release 99 specifications. This needs to be signaled at the bearer establishment phase. It is also necessary for the network to be aware that the call is a video telephony call. The tariff used to bill for video telephony calls may be different to that for speech calls, and the network may have to handle and route video calls differently to speech calls in some circumstances, for example when handling a request to establish a call between a mobile customer using a 3G-324M terminal and a different type of terminal, perhaps on another network.

Specification of the call setup procedures is outside the scope of 3G-324M. Setup of the bearer can be split into the signaling between the UE and the MSC, which uses Q.931 like signaling; and signaling between the MSCs, which uses SS7. The setup messages for a call between two 3G-324M mobile terminals are outlined in 3GPP documents TS 24.008 "mobile radio interface layer 3 specification; core network protocols—stage 3," and TS 27.001 "general on terminal adaptation functions (TAF) for mobile stations (MS)."

To ensure that a bearer of the correct type is established some of the *bearer capability information elements* (BC IEs) are set to indicate this. The *information transfer capability* (ITC) must be signaled as *unrestricted digital information* (UDI) call (ITC = UDI) as opposed to a speech call. The duplex mode should be set to full duplex and sync/async set to synchronous. The required channel must be signaled as transparent by setting the *connection element* to transparent (CE = T). The *wanted air interface user rate* (WAIUR) and *fixed network user rate* (FNUR) should both be set to 64 kbit/s. It must also indicate that H.223 and H.245 is required by setting the user information layer 1 to H.223&245 and setting bits 1–5 of the *low layer compatibility* (LLC) octet 5 to binary 00110, so that the network and called terminal are aware that the call is a multimedia call rather than a speech call. One of the most common reasons for video calls failing is because not all the bearer capability information has propagated end to end.

4.9 Chapter Summary

We have looked at the components of a video telephony terminal that are within the scope of 3G-324M and the role each of them plays. Together with the necessary bearer information elements in the call signaling these components are used to set up video telephony calls by going through the following roughly sequential steps to establish a multimedia session:

1. Set up bearer through signaling of H.223/H.245.
2. H.223 multiplexers agree a common mobile level.
3. H.245 using LCN0 and MTE0 performs
   a. Terminal capability exchange
   b. Master/slave determination
   c. Opening of logical channels for media information
   d. Exchange of multiplex table entries, allowing media information to be multiplexed and demultiplexed
4. Logical channels are used to send and receive media information.

Steps c and d may be performed in reverse order.

Chapter 5

The H.223 Multiplexer in Detail

To understand 3G-324M in more depth, we have to look at H.223 in detail. We have seen that the *multiplexer* is a two-layer structure. In this chapter we will discuss bit and octet ordering in H.223, review adaptation layer types 1–3, and examine the structure of MUX-PDUs for mobile levels 0, 1, and 2 in more detail. The specification of multiplex table entries will be described and we will look at how the multiplexer handles control, audio, video, and data information.

5.1 Bit and Octet Ordering

H.223 MUX-PDUs are transmitted as a series of bits. Referring to Figure 5.2, the *least significant bit* (LSB) of the lowest numbered octet is transmitted first and the *most significant bit* (MSB) of the highest numbered octet is transmitted last. In general the contents of the fields specified in the MUX-PDU are also laid out with their LSB corresponding to the lowest bit of the lowest numbered octet and their MSB corresponding to the highest bit of the highest numbered octet allocated to the field. CRCs are an exception to this rule, they are all mapped into octets of the MUX-PDU in the opposite way; their MSB is placed in the lowest bit of the lowest numbered octet and their LSB in the highest bit of the highest numbered octet allocated to them.

Some processes, including H.263 video codecs, G.723.1 audio codecs, and H.245 provide AL-SDUs to the multiplexer as octets that are organized MSB first. This is not a problem, because the multiplexer is not aware of the internal structure of this data, and delivers it transparently from end to end: the bits and octets come out of the far-end multiplexer in the same order that they went into the local multiplexer.

5.2   Adaptation Layer Formats

The format of AL-PDUs produced by AL1, AL2, and AL3 is shown in Figure 5.1. All AL-SDUs must be a whole number of octets, as must the AL-PDUs produced by the adaptation layer. The adaptation layer type to be used for a particular media or control information stream is specified when a logical channel is opened, using H.245.

5.2.1   Adaptation layer type 1

The format of an AL-PDU for AL1 can be seen in Figure 5.1. No additional header or CRC information is added to the incoming AL-SDU, which is passed through transparently. AL1 can operate in two modes: framed and unframed—the format of the AL-PDU is the same for both modes. In framed mode AL1 takes each AL-SDU and passes it

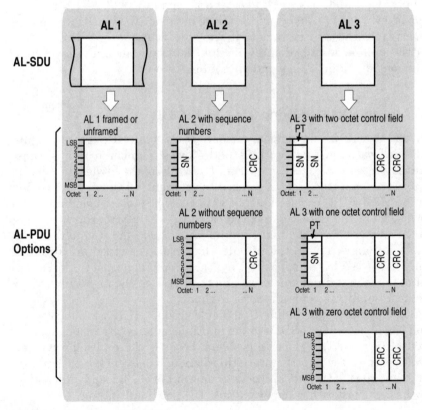

Figure 5.1   Structure of AL-PDUs.

unchanged to the MUX layer. In unframed mode, incoming AL-SDUs can be broken up and passed to the MUX layer as multiple AL-PDUs.

The unframed mode is intended for information streams that have no structure and can be considered to be a single AL-SDU of indefinite length. They are therefore segmented at the adaptation layer (other segmentable logical channels allow segmentation of AL-PDUs/MUX-SDUs by the multiplexer layer). MUX-PDUs that carry any AL-SDUs for a logical channel that is using unframed AL1 should never indicate that they contain the last segment, so the PM flag should never be set (or in the case of ML2, the closing flag should never be 1s complemented).

5.2.2   Adaptation layer type 2

As shown in Figure 5.2, AL2 appends a 1-octet CRC to the incoming AL-SDU. This provides a means for detecting errors in the received AL-PDU. Optionally AL2 may also add a 1-octet header representing a modulo 256 sequence number that is incremented with every AL-PDU formed.

The CRC is calculated by treating the whole of the AL-PDU, including any sequence number but excluding the CRC itself, as a binary number B, with bit 1 of the first octet as the MSB and bit 8 of the last octet before the CRC as the LSB. B is multiplied by binary 10000000 and then divided modulo 2 by binary 10000111. The 8-bit remainder is bit reversed to give the CRC (the MSB of the remainder is the LSB of the CRC).

5.2.3   Adaptation layer type 3

Possible formats of AL-PDUs for AL3 are shown in Figure. 5.1. An optional 1- or 2-octet control field header may be added to the AL-SDU and a 2-octet CRC is appended.

The CRC is calculated by treating the whole of the AL-PDU, including any control field that is present but excluding the CRC itself, as a binary number B, with bit 1 of the first octet as the MSB and bit 8 of the last octet before the CRC as the LSB. The number of bits in B (the length of B) is represented by the number k. B is multiplied by binary 1000000000000000 (2^{16}) and then divided modulo 2 by binary 1000100000010001. The remainder is added to the remainder generated by multiplying 2^k by binary 1111111111111101 and then dividing modulo 2 by 1000100000010001. The result is 1's complemented and bit-reversed to give the CRC (the MSB of the remainder is the LSB of the CRC).

The control field comprises a single bit *payload type* (PT) field in bit 1 of the first octet of the field, followed by a 7- or 15-bit sequence number. The payload type field indicates that the AL-PDU is sending an AL-SDU when PT is set to one (1) and is known as an I-PDU (information PDU).

When PT is set to zero (0) the AL-PDU contains a single-octet message related to the retransmission procedure and is known as an S-PDU (*supervisory PDU*). For S-PDUs, the sequence number field contains the sequence number of the I-PDU that the retransmission request relates to. There are two possible values for the single-octet payload message of an S-PDU: 0 0x00 *selective reject* (SREJ) is used by the receiver to request retransmission of an I-PDU, 0xFF *declined retransmission* (DTRX) is used by the transmitter to refuse the SREJ retransmission request.

The use of the optional retransmission procedure of AL3 means that the transmitter must buffer the I-PDUs that it sends, in case a request to retransmit is received. If a SREJ request is received for an I-PDU that is still in the buffer, it can be retransmitted. If the I-PDU is no longer in the buffer the terminal must send a DTRX message. The receiving terminal must also buffer incoming data, increasing the end-to-end delay, if it is to be able to make any use of retransmitted I-PDUs.

5.3   Interfacing to the Adaptation Layer

The process in a multimedia terminal that exchanges AL-SDUs with the multiplexer is referred to in the H.223 recommendation as the ALx user, where x may be 1, 2, or 3 (or ALxM, if mobile level 3 is used).

The adaptation layer interacts with the AL user through a set of primitives. The primitives common to all three adaptation-layer types are AL-**DATA.request(AL-SDU)**, which is used to transfer an outgoing AL-SDU to the multiplexer, and **AL-DATA.indication(AL-SDU, EI)**, which the multiplexer uses to transfer an incoming AL-SDU to the user, along with any *error indication* (EI), which may have resulted from a failure of the local CRC check. The EI parameter is not used with AL1 because it does not include a CRC. Other primitives specified in H.223 are associated with the facility to abort delivery of partially delivered AL-SDUs. These are not described here because the 3G-324M *terminal implementers guide* (TR 26.911 "codec(s) for circuit-switched multimedia telephony service; terminal implementers guide (Release 1999)") says that the H.223 abort procedures should not be used.

AL3 has an associated primitive, **AL-DTRX.indication**, which is used in the optional retransmission procedure that AL3 supports. **AL-DTRX.indication** is used to indicate to the AL3 user that the local AL3 has sent an S-PDU containing a DTRX message, declining an SREJ request received from the remote terminal to retransmit an I-PDU. This is of value where the process supplying the AL-SDUs can modify its behavior in a useful way as a result of receiving the indication. For instance, a video codec using AL3 could send an *intra* coded frame if this message is received.

The coding of these primitives and the definition of different error indications is outside of the scope of the H.223 recommendation and is left to the implementer of the terminal.

5.4 MUX-PDU Formats

The format of MUX-PDUs produced by ML0, ML1, and ML2 is shown in Figure 5.2. All MUX-PDUs are a whole number of octets. The mobile level to be used during a multimedia session is determined by the exchange of flags during the initialization process described in Chapter 4. Mobile levels can be changed during a session, using H.245 messages.

5.4.1 Mobile level 0

For ML0, MUX-PDUs are delineated by the single-octet flag 0x7E (binary 01111110) as the opening and closing flag. The closing flag can also be the opening flag for the next MUX-PDU. When the multiplexer has no completed MUX-PDUs to send, it sends repeated 0x7E flags as stuffing.

The first octet of a ML0 MUX-PDU is a header. This is divided as shown in Figure 5.2. The LSB contains the *packet marker* (PM) bit. If this bit is set to 1 it indicates that the previous MUX-PDU contained the last MUX-SDU for a segmentable logical channel. This means that only one segmentable logical channel can have the final segment of a MUX-SDU/AL-PDU in any MUX-PDU. The last octet of the final segment of the MUX-SDU for the logical channel that is being indicated by the PM bit occupies the last octet of the MUX-PDU, immediately before the closing flag.

The next 4 bits of the header represent the MC, which specifies the MTE being used for this MUX-PDU. The three most significant bits are the HEC field containing the CRC, to provide a way of detecting whether an error has occurred in the transmission of the header information.

The CRC for the HEC is computed by taking the MC field and bit-reversing it to form a 4-bit binary number B so that bit 2 (the LSB of the MC field) is the MSB of B and bit 5 is the LSB of B. B is multiplied by binary 1000 and divided modulo 2 by 1011. The remainder is bit-reversed to form the CRC for the HEC. The example in Chapter 2 illustrating how CRCs are computed was based on the computation of the HEC field. Because there are only 16 possible values for MC, the implementation of this function for forming MUX-PDU headers and for checking the CRC of received MUX-PDUs can be based on storing precomputed HEC CRC values in a lookup table. Any received MUX-PDU that fails the CRC check should be discarded.

In ML0 operation, to avoid flag emulation within the body of the MUX-PDU the multiplexer examines the payload prior to transmission

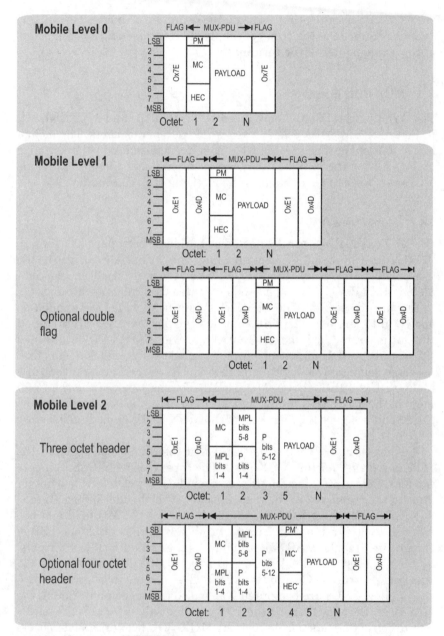

Figure 5.2 Structure of MUX-PDUs.

and inserts a "0" after each occurrence of five contiguous "1s". The receiving demultiplexer removes any "0" after five contiguous "1s" in the received MUX-PDUs prior to further processing.

ML0 corresponds to the operation of H.324 without Annex C, intended for operation over the GSTN using a synchronous modem. For modem-based operation there is no underlying octet framing structure at the network layer, so the "0" bit insertion technique does not need to worry about causing misalignment of MUX-PDU octets with any underlying octet framing. For the 3G UMTS network there is an underlying octet ori-entated structure specified in 3GPP document TS 25.301 "radio interface protocol architecture." If octets of the MUX-PDU are aligned with the octets of the higher layer PDU specified in this document, this can be exploited to reduce computational requirements by making the informa-tion easier to extract from the incoming bit stream. If the 0-bit insertion technique is used, the alignment of MUX-PDU octets with any under-lying network octet framing structure can no longer be guaranteed. This means that for ML0 flag detection must be performed at all possi-ble bit positions. When the flags are found, the multiplexer then exam-ines the data between the flags and removes every 0 that follows five consecutive 1s to extract the MUX-PDU.

5.4.2   Mobile level 1

As shown in Figure 5.2, MUX-PDUs for ML1 are identical to those for ML0, except for the flags used to delimit them. A 2-octet opening and closing flag, 0xE14D, is mandatory. Optionally this can be repeated to form a double flag. The header structure for the MUX-PDU is the same as for ML0. Stuffing is performed by repeating the flag, so if the double-flag option is used any stuffing will be an even number of 0xE14D values. ML1 cannot be started in double-flag mode. If a terminal sup-ports the option of double flags it must always start using single 2-octet flags, and can then be switched to double flag operation via an H.245 command.

MUX-PDUs are extracted from the incoming bit stream by correlat-ing it with the 0xE14D flag. There is no procedure for avoiding flag emulation in the payload in ML1: the longer flag and octet alignment with the underlying network framing are expected to significantly reduce the problem of flag emulation—sequences that emulate flags but that are not octet aligned need not be considered. Where flag emu-lation does result in a false start to a MUX-PDU being detected, it can be eliminated with high probability by examining the header octet to see if the CRC in the HEC field is consistent with the contents of the MC field. This two-stage verification should result in problems with flag emulation being negligible.

5.4.3   Mobile level 2

The structure of MUX-PDUs for ML2 is shown in Figure 5.2. The opening and closing flag is the same as for ML1, but there is no double flag option. The header field is 3 octets or optionally 4 octets long. In both cases the first 3 octets contain, in increasing order of bit and octet significance, the 4-bit MC field with the same meaning as for ML0 and ML1, the 8-bit *Multiplex Payload Length* (MPL) field, and a 12-bit parity check field for the header, which replaces the HEC field used in ML0 and ML1.

The parity check field is generated using the extended Golay (24, 12, 8) code 110001110101. The computation is similar to that of a CRC: the MC and MPL fields are treated as a single binary number, multiplied by 2^{11} and divided modulo 2 by the bit-reversed extended Golay code. The remainder, with a parity bit added (1 if an even number of 1s, 0 if odd) in the LSB position, is bit reversed to provide the contents of the parity check field. When a ML1 MUX-PDU is received the entire header is again divided modulo 2 by the extended Golay code to produce a 12-bit remainder known as the syndrome. If this is zero then no errors occurred during transmission of the MUX-PDU. If small numbers of bit errors (<3) occur, they can be detected and corrected by examining the syndrome, which will have 1s at the bit positions corresponding to the position in the combined MC+MPL fields where the errors have occurred. Higher numbers of errors may be detected and corrected by more complex calculations using the syndrome.

The 8-bit MPL implies that MUX-PDUs must not contain more than 254 octets of payload information (255 or 0xFF is a reserved value).

At ML2 there is no PM flag to indicate that the previous MUX-PDU contained the last segment of a segmentable logical channel. This is indicated by 1s complementing the closing flag rather than using a PM bit. This must be reflected in the procedure for detecting MUX-PDU boundaries. As for ML1, boundaries are detected by correlation. The 0xE14D flag has as many 0s as 1s in it. By treating 0s in the MUX-PDU as having the value −1 when performing the correlation, boundaries can be detected by looking for both positive and negative peaks in the correlation result.

The optional 4-octet header for ML2 is also shown in Figure 5.2. The last octet of the header is made up of a PM, MC, and HEC field for the previous MUX-PDU, computed as described for ML0 and ML1. This allows cross-checking, further improving robustness to higher error rates and reliability in extracting MUX-PDUs under such conditions. ML2 cannot be started in 4-octet header mode. If a terminal supports this option, it must always start (following multiplexer synchronization at ML2) using the mandatory 3-octet header, and can then be switched to ML2 with optional header operation via an H.245 command.

When there is no completed MUX-PDU available to send, ML2 sends empty MUX-PDUs with the MC and MPL fields set to all zeros (and therefore the parity bits become all zero). This means that when using a 3-octet header the stuffing sequence will be a repetition of E1 4D 00 00 00.

5.5 MUX Table Entries

Having looked at the framing of MUX-PDUs, we will now look at how the format of the payload or information field is specified. The payload is constructed from MUX-SDUs (AL-PDUs) delivered to the multiplexer layer from the adaptation layer. These are combined to form MUX-PDUs according to "recipes" known as MTEs. Each MTE has an associated number, the MC, which is included in the header of each MUX-PDU. The recipes and their associated MCs are signaled to the remote terminal during multimedia session establishment.

5.5.1 Specifying MTEs

MTEs are described by allocating each octet in the payload to a particular logical channel. A payload can include information from multiple logical channels. In the case of nonsegmentable logical channels the MUX-SDU must be placed in a contiguous set of octets within a single MUX-PDU. MUX-SDUs for segmentable logical channels can be broken into segments and assigned to noncontiguous octets within a MUX-PDU, or to multiple MUX-PDUs. As we have seen, when the last segment of a MUX-SDU is placed in a MUX-PDU this is signaled to the remote terminal using the PM bit in the header of the following MUX-PDU (or by 1s complementing the closing flag if ML2 is being used) so it is only possible to place the last segment of one segmentable logical channel in each MUX-PDU. This will be identified as the logical channel occupying the octet immediately before the closing flag.

MultiplexTableEntryDescriptors are used to describe how the octets of the information field are allocated. The syntax of a descriptor is given by MultiplexTableEntryDescriptor = {MultiplexElement}, {MultiplexElement}, ... , {MultiplexElement}, where MultiplexElement = {Element, Repeat Count}. An Element may either be a specific logical channel number (e.g., LCN1) or a subelement list with the same format as a MultiplexTableEntryDescriptor, leading to the ability to produce complex nested descriptions of MTEs. The repeat count (RC) is either finite, specifying the length of the slot as an integer number of octets (e.g., RC20), or it will specify that it continues indefinitely until the closing flag (RC UCF). RC UCF must be used for the last MultiplexElement in the MultiplexTableEntryDescriptor, and only for the last MultiplexElement.

A simple example of a MultiplexTableEntryDescriptor is {LCN1, RC24}, {LCN2, RC UCF}, which specifies that the first 24 octets of the MUX-PDU information field will contain information from LCN1, and the rest will contain information from LCN2.

The definition of MultiplexTableEntryDescriptor is recursive, because a MultiplexElement can contain a MultiplexTableEntryDescriptor. This means that nested definitions are possible, such as {{LCN1, RC8}, {LCN2, RC8}}, RC UCF}, which specifies that the MUX-PDU information field contains a repeating pattern of 8 octets of LCN1 alternating with 8 octets of LCN2 until the closing flag. A further example, {LCN1, RC4}, {{LCN2, RC8}, {LCN3, RC10}, RC UCF} specifies that the MUX-PDU information field contains a pattern of 4 octets of LCN1 followed by 8 octets of LCN2, alternating with 8 octets of LCN3 that repeats until the closing flag.

We can classify the MultiplexTableEntryDescriptor {{LCN1, RC8}, {LCN2, RC8}}, RC UCF} as having a top-level element list size of 1, a subelement list size of 2 and a nesting depth of 1. MultiplexTableEntryDescriptor {LCN1, RC4}, {{LCN2, RC8}, {LCN3, RC10}, RC UCF} has an element list size of 2, a subelement list size of 2, and a nesting depth of 1. The simpler MultiplexTableEntryDescriptor {LCN1, RC24}, {LCN2, RC UCF} has an element list size of 2, a subelement list size of 0, and a nesting depth of 0.

H.223 does not allow unlimited element list size, subelement list size, or nesting depth. In the basic mode of operation, the multiplexer must support an element list size of up to 2, a subelement list size of up to 2, and a nesting depth of up to 1. Optionally the multiplexer may be able to operate in an enhanced mode, where it can support an element list size of up to 255, a subelement list size of up to 255, and a nesting depth of up to 15. In practice such large list sizes and nesting depths are unlikely to be useful.

Whether the multiplexer supports basic or enhanced h223MultiplexTableCapability is declared as part of the capability exchange when the session is being established and refers to its capability to interpret received MUX-PDUs. If enhanced capability is declared, the maximum element list size, subelement list size, and nesting depth supported must also be explicitly declared. Its MultiplexTableEntryDescriptors, describing how it will construct and send MUX-PDUs, are sent to the remote terminal as part of the MultiplexTableEntryExchange when the session is being established.

A terminal may well support the ability to receive and interpret MUX-PDUs formed using enhanced MultiplexTableEntryDescriptors but only use basic MultiplexTableEntryDescriptors to form MUX-PDUs for transmission. The transmitting terminal must respect the capabilities of the receiver. Up to 15 MultiplexTableEntryDescriptors are allowed, representing MTE1–MTE15. MTE0 is predefined for use with LCN0 to establish the session—its (implicit) MultiplexTableEntryDescriptor is {LCN0, RC UCF}.

5.5.2 Constructing MUX-PDUs

The payload of a MUX-PDU is formatted using the MTE referenced by the MC in its header. The payload of the MUX-PDU can be ended at any point by the closing flag. The resulting MUX-PDU may be a subset of the structure described by the MultiplexTableEntryDescriptor; not every octet needs to be present provided that those that are present accord with the MultiplexTableEntryDescriptor.

MUX-PDUs are made up of information from incoming MUX-SDUs. For nonsegmentable logical channels the MUX-SDUs cannot be broken up: a MUX-PDU must contain a whole MUX-SDU mapped into the octets specified by the MultiplexTableEntryDescriptor for the associated LCN. This means that the size of the MUX-SDU must either match the specified repeat count or be less than this, in which case the MUX-PDU must be immediately terminated by a closing flag.

Figure 5.3 shows several different MUX-PDUs, sent one immediately after the other using ML1, constructed using the same MultiplexTableEntryDescriptor {LCN1 RC4}, {LCN2, RC UCF}, which has been assigned to MTE1. In this example LCN1 is a nonsegmentable logical channel and LCN2 is segmentable. The MUX-PDUs are preceded by the 2-octet flag 0xE14D. The opening flag of a subsequent MUX-PDU is also the closing flag of the previous MUX-PDU. The first octet of the MUX-PDUs is the header, with MC = 0001 and HEC =101. PM is set to 1 for the first MUX-PDU, giving a header value of 101 0001 1 = 0xA3, indicating that the LCN2 data in the previous MUX-PDU (not shown) was the last segment of an LCN2 MUX-SDU for that channel. In the first MUX-PDU the information field comprises a 4-octet LCN1 MUX-SDU followed by the first 6 octets of an LCN2 MUX-SDU. For the second MUX-PDU the PM flag is set to 0, giving a header value of 0xA2. The closing flag occurs sooner than for the first MUX-PDU, after a 4-octet LCN1 MUX-SDU and 2 octets of an LCN2 MUX-SDU have been included. The final MUX-PDU of Figure 5.3 has the PM bit set to 1 giving a header value of 0xA3, indicating that the LCN2 information in the previous MUX-PDU was the last segment of an LCN2 MUX-SDU for that channel. The final MUX-PDU contains a 2-octet LCN1 MUX-SDU, terminated by the closing flag. So these three MUX-PDUs contain one 10-octet LCN2 MUX-SDU segmented over two MUX-PDUs, and three LCN1 MUX-SDUs, of 4, 4, and 2 octets.

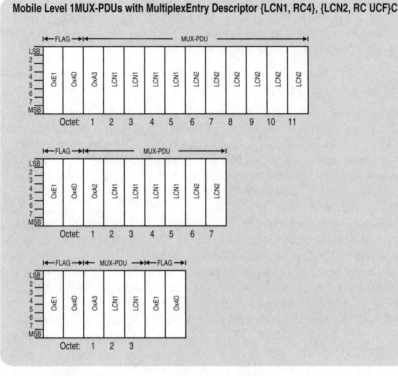

Figure 5.3 MUX-PDUs created using an example MTE.

5.5.3 Designing MTEs and building MUX-PDUs

The definition of MultiplexTableEntryDescriptors for a H.324, H.324M, or 3G-324M terminal is left up to the implementer, as is the method of dynamically building MUX-PDUs for sending. A knowledge of the likely sizes of MUX-SDUs that are presented to the multiplex layer helps in the design of useful MultiplexTableEntryDescriptors for MUX-PDUs.

There are restrictions on the combinations of logical channels that can be defined by a MultiplexTableEntryDescriptor and carried within a single MUX-PDU. A MUX-PDU can contain multiple MUX-SDUs from a single nonsegmentable logical channel; however, this is only possible if the receiving terminal declares support for enhanced multiplex capability.

In the 3G-324M implementer's guide it is recommended that MUX-PDUs be limited to less than 200 octets, to reduce the impact of transmission errors, and that terminals support the H.245 maxPDUSize-Capability, allowing them to respond to requests to limit the size of the MUX-PDUs they send.

5.6   Inputs to the Multiplexer

The processes that use the multiplexer use the features it offers in different ways. We review here the ways that different types of information use the features of the multiplexer.

5.6.1   Control

Control messages use LCN0, a segmentable logical channel that is implicitly open from the start of a multimedia session. The output of the H.245 procedure is MultimediaSystemControlPDU messages, containing one or more H.245 ASN.1 control messages, formed into binary messages using PER. These are broken into CCSRL-PDUs and then formed into NSRP command frames, which represent the AL-SDUs that make use of AL1 in framed mode to form AL-PDUs/MUX-SDUs.

At the start of a session, the control MUX-SDUs are formed into MUX-PDUs based on MTE0, the only predefined multiplex table entry. Once other MTEs have been exchanged these can also be used to send control information, provided that the MultiplexEntryDescriptor includes support for LCN0 MUX-SDUs. One possible beneficial way of exploiting this is to provide some MTEs with the capability to carry an NSRP response frame, consisting of 4 octets, in addition to logical channels for media.

5.6.2   Audio

Audio is made up of regular small MUX-SDUs that are carried in a nonsegmentable logical channel. AL2 is designed to carry audio information. H.324 does not recommend using AL2 with sequence numbers for G.723.1 because it considers that they are not useful given that the jitter (difference in arrival times of AL-PDUs) on a circuit-switched network is likely to be less than the time between frames of audio data. This is also likely to be true for GSM-AMR.

A G.723.1 codec produces frames of either 24 octets or 20 octets every 30 ms, depending on whether it is operating in the 6.3-kbit/s high rate or 5.3-kbit/s low rate mode. If AL2 without sequence numbers is used, the addition of a CRC octet results in AL-PDUs that are either 25 or 21 octets long and SID frames that are 5 octets long. This can be used in the design of MTEs to ensure that there is a selection of matching repeat counts for audio logical channels.

The same approach can be taken to the design of MTEs for GSM-AMR. The eight modes of GSM-AMR: 12.2, 10.2, 7.95, 7.4, 6.7, 5.9, 5.15, and 4.75 kbit/s produce AL-PDUs of 32, 27, 22, 20, 19, 17, 15, and 14 octets respectively, and SID frames will be 7 octets in size after addition of the 1-octet CRC.

3G-324M recommends that no more than three audio frames should be included in any single MUX-PDU, to keep audio delay low.

5.6.3 Video

Video is carried using a segmentable logical channel. AL3, with its capability to request retransmission of AL-PDUs/MUX-SDUs, is designed for transporting video. The provision of a retransmission capability implies that a logical channel is set up that is not unidirectional, but which has the capability to send AL-PDUs in both directions (I-PDUs in the forward direction and S-PDUs in the reverse). To do this requires that a bidirectional logical channel be set up. As we will see in the next chapter H.245 provides the ability to open bidirectional logical channels that comprise two associated unidirectional logical channels.

If video is to be established in both directions at AL3, a single bidirectional logical channel should be used. To avoid the problems that might arise if both terminals try to simultaneously open bidirectional logical channels, H.245 defines a procedure that makes use of the master/slave status of the terminals to resolve the conflict.

H.324 recommends that if a DTRX message is received from the remote terminal, the video codec should encode the next frame as an I-frame; however, 3GPP recommends that the retransmission procedure is not used for video over AL3 due to the delay that results from having to maintain buffers.

Video need not be restricted to using AL3. The 3G-324M implementer's guide recommends that video over AL2 is supported in addition to AL3. It implies that this may be preferable, particularly as the retransmission procedure of AL3 is not recommended and AL2 has lower overheads. In this case unrelated unidirectional logical channels are used to transport the video in either direction.

H.324 requires that PSCs for H.263 encoded video are aligned with the start of an AL-SDU. 3G-324M further recommends that AL-SDUs for video that do not contain a PSC should start with a GOB (or by implication slice) header to improve error resilience.

For MPEG4, 3G-324M requires a similar approach to that taken for H.263, by ensuring that start codes and headers are aligned with the start of AL-SDUs.

Although H.261 is an optional codec for 3G-324M, it is not discussed here as it is unlikely to be widely used in practice.

5.6.4 Data

Little has been said so far about supporting data. Logical channels can be opened to support data applications and typically would use AL1, with the data application itself providing any error protection mechanisms that are needed.

The H.324 recommendation discusses several data applications that can be used, for such things as application sharing, image and file transfer, and real-time control. The best known of these is ITU-T T.120 "data protocols for multimedia data conferencing." The T.120 stack is large and may be overkill for a video telephony handset. Transmitting and receiving data whilst maintaining audio and video communication is limited by the available bandwidth on a 64-kbit link. 3G-324M video telephones will also have access to packet-based services via the SGSN, so in many cases this will be the best way to provide access to data applications. An ancillary capability to transmit and receive data that is closely coupled to the video telephony call may be required in some circumstances. Examples may include interacting with a video mail server or some other supplementary service. In such cases the amounts of data that will be transferred are small, and can be handled in several possible ways by H.245 messages such as `UserInputIndication` in LCN0. This will be discussed in Chapter 6.

5.7   Chapter Summary

We have looked in more depth at the H.223 multiplexer. Its two-layer structure, and the use and function of each layer have been described.

The adaptation layer provides three alternatives for generating AL-PDUs, with different levels of error protection that can be applied independently to individual information streams. The multiplexer has three modes of operation: mobile levels 0, 1, and 2. These modes of operation provide different levels of protection for the MUX-PDU as a whole and are intended to be used in different transmission error conditions.

We have seen how the format of MUX-PDU payloads can be described using multiplex table entries.

The use of the multiplexer to handle different media types and control information illustrated the appropriate use of segmentable and nonsegmentable logical channels, and gave some indications of design considerations when constructing MTEs.

After reading this chapter you should have a good understanding of the upper adaptation layer of the multiplexer and the structure of AL-PDUs. The structure of MUX-PDUs at mobile levels 0, 1, and 2 and the use of MTEs to specify the allocation of octets in the payload or information field should also be clear.

6

H.245 Command and Control in Detail

H.245 is a recommendation specified for the control component of several ITU-T multimedia terminal recommendations. The H.245 document is large, covering the procedures to be used and the syntax and semantics of the messages that are exchanged in all these terminal standards. We will restrict our discussion of H.245 to its use in 3G-324M, but even so a comprehensive description of all the features of the recommendation is not possible. In this chapter we will go into some depth on how H.245 messages are constructed and used, but ultimately anyone undertaking development of terminals or systems for 3G-324M should refer to the source documents.

One of the purposes of H.245 is to provide a way of informing the remote terminal of the capabilities and limitations of the local terminal. This includes the features of the multiplexer and also the different media formats supported. It is possible that a terminal may be able to support a number of alternative media codecs, but have limitations on the combinations of codecs that it can operate simultaneously, due perhaps to available resources. How this is communicated using the capability exchange procedure will be described.

The procedures of capability exchange, master slave determination, opening logical channels, and multiplex table entry exchange were introduced in Chapter 4. In this chapter we will look at each of these procedures and their associated signaling entities in more detail, and describe the additional procedures that H.245 provides.

Once a session has been established, additional commands and indications may be used for a variety of purposes. The function and purpose of the most important commands and indications will be reviewed.

H.245 provides specific "hooks" or code points for some video and audio codecs. Other codecs can be specified within H.245 messages by defining and using generic capabilities; the final section of the chapter describes how this is done.

6.1 H.245 Messages

This chapter is intended to be read in conjunction with Appendix A, which provides a summary of the syntax of those H.245 MultimediaSystemControlMessages most commonly used in mobile video telephony for easy reference. It must be emphasized that Appendix A represents a subset of the full set of H.245 messages, which should cover most of the messages used in 3G-324M video telephony, but may not cover all. The message structures in Appendix A, as in the H.245 recommendation itself, are defined hierarchically. The first section of the Appendix shows that MultimediaSystemControlMessages can be classified as being request messages, response messages, commands, or indications. It then provides a list of messages in each of these categories. Subsequent sections show the structure of the messages used in each

TABLE 6.1   H.245 Procedures

Procedure	Associated SEs	Instances	Persistent
Master/slave determination	MSDSE	One	No
Capability exchange	Outgoing CESE Incoming CESE	One	No
Unidirectional logical channel signaling	Outgoing LCSE Incoming LCSE	One per unidirectional logical channel	Yes
Bidirectional logical channel signaling	Outgoing B-LCSE Incoming B-LCSE	One per bidirectional logical channel	Yes
Close logical channel	Outgoing CLCSE Incoming CLCSE	One per logical channel	No
Multiplex entry exchange	Outgoing MTSE Incoming MTSE	One per MTE	No
Request multiplex entry	Outgoing RMESE Incoming RMESE	One per MTE	No
Mode request	Outgoing MRSE Incoming MRSE	One	No
Round trip delay	RTDSE	One	No
Maintenance loop	Outgoing MLSE Incoming MLSE	One per bidirectional logical channel	Yes

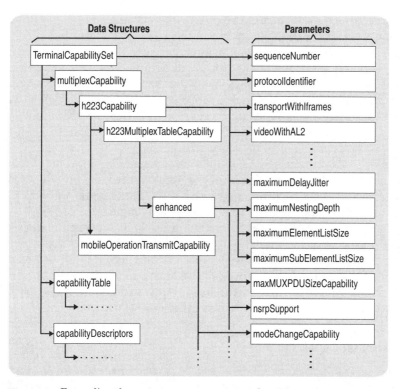

Figure 6.1 Expanding the `multiplexCapability` data structure.

of the procedures of H.245 shown in Table 6.1, and then list the syntax of command and indication messages that are not part of any signaling entity but may be used in a multimedia session.

H.245 messages vary in complexity. They are data structures containing fields, which themselves may either be data structures or parameters with an associated type such as INTEGER, BOOLEAN, and the like. To illustrate this, Figure 6.1 provides a graphical illustration of how the `TerminalCapabilitySet` message in Appendix A expands, using the hierarchy of definitions within the Appendix, to create an instance of the message. The `TerminalCapabilitySet` message contains `sequenceNumber`, `protocolIdentifier`, `multiplexCapability`, `capabilityTable`, and `capabilityDescriptors` fields. The first two of these are parameters, the last three are themselves data structures.

The `multiplexCapability` data structure offers a choice. In the H.245 recommendation this choice includes `H222Capability`, `H2250Capability`, and other choices relating to the use of H.245 in recommendations other than H.324. In Appendix A the list of choices has only one entry (no choice!)—`H223Capability`. The `H223Capability` data structure contains a mixture of parameters and data structures. Eventually, as all the data structures of the `TerminalCapabilitySet` message (or in general any H.245 message) are expanded and filled out, the message is created as a structured list of parameter values separated by identifiers for the relevant data structure that the parameters belong to. The PER encoding process then compresses this into a highly efficient binary representation by, for example, representing BOOLEANs as single bits and using short tags to code the data structure (and length values if the data structure is of variable length).

When interpreting the syntax definitions of Appendix A, it is helpful to know that `CHOICE` means only one of the items in the list that follows is included in the message to be constructed, `SEQUENCE` means all items in the list must be included unless they are explicitly marked as `OPTIONAL`. `SET SIZE (1...N) OF item` means that multiple instances can be present, within the range specified.

It must be emphasized that Appendix A is only suitable for use as a reference for looking at the syntax of the H.245 messages used in 3G-324M; the complete syntax of H.245 is required by general purpose ASN.1 compilers that produce PER or BER messages.

6.2   Terminal Capabilities

The `capabilityTable` data structure within the `Terminal CapabilitySet` message is used to list the audio and video codecs and other capabilities that the terminal supports, and to describe which modes of operation and algorithm options are supported within the relevant audio or video standard.

The `capabilityTable` is a set of `capabilityTableEntries`, a numbered list of capabilities as shown in Figure 6.2. Each `capability` describes the capability of the terminal to support a particular video or audio coding format. It may also describe the capability to transmit and receive user inputs via any keyboard or keypad. Other possibilities are defined in the H.245 recommendation, including the capability to support data applications, but we will not consider them here. Each `capability` can be declared as a receive, transmit, or receive-and-transmit capability—allowing the terminal to declare whether it relates to its ability to receive or transmit the media type specified or whether there is a limitation that it must transmit and receive the same media format. For instance if the terminal can support H.263 and MPEG4 video, but in either case it must use one format to both receive and transmit, this limitation can be explicitly stated. Transmit capabilities can be used by the terminal to signal preferences to the receiver—the ultimate limitation on the transmitting terminal is that it must respect the receive capabilities of the remote terminal.

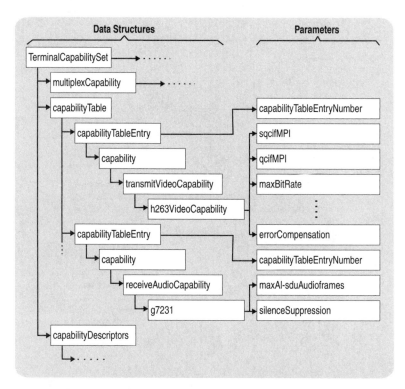

Figure 6.2 Expanding the capabilityTable data structure.

A terminal must declare its receiving capabilities. It need not declare its transmitting capabilities, provided that it respects the declared capabilities of the remote terminal.

Having declared all the individual capabilities it supports through a list of capabilityTableEntry structures, the allowed combinations of these capabilities are declared using a set of CapabilityDescriptors as shown in Figure 6.3. Each numbered capabilityDescriptor describes the simultaneousCapabilities that the terminal supports. These are made up of a set of alternativeCapabilitySet structures. The terminal can simultaneously support only one of the capabilities referenced by the capabilityTableEntryNumbers within each alternativeCapabilitySet. An example should make this clearer. Assume we have a hypothetical terminal that is capable of supporting a single logical channel for video, which can be in either H.263 or MPEG format, and a logical channel for audio that can be in either G.723.1 or GSM-AMR format. It can alternatively support two simultaneous logical channels for video, provided that they are both in MPEG4 format and

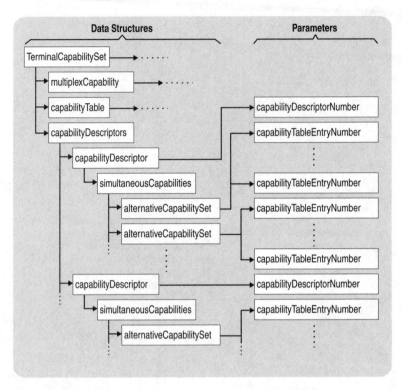

Figure 6.3 Expanding the `capabilityDescriptors` data structure.

that any logical channel for audio is restricted to GSM-AMR. To do this we need to have two `capabilityDescriptor` structures as shown in Figure 6.4, describing two `simultaneousCapabilities`. The first is made up of two `alternativeCapabilitySet` structures, and the second is made up of three.

A partial listing of the ASN.1 for a `TerminalCapabilitySet` request message for the capability set example we have just described is shown in Figure 6.5.

Referring to Figures 6.4 and 6.5, a terminal receiving this `terminal CapabilitySet` can either choose to use capability descriptor 1 or capability descriptor 2 when deciding which logical channels to open. If it uses capability descriptor 1, it can open a single audio logical channel that is either GSM-AMR or G.723.1 and a single video channel that is either H.263 or MPEG4. If it uses capability descriptor 2, it can open three logical channels, but there is no choice of media within each of these channels—one must be GSM-AMR, the other two are both limited to MPEG4.

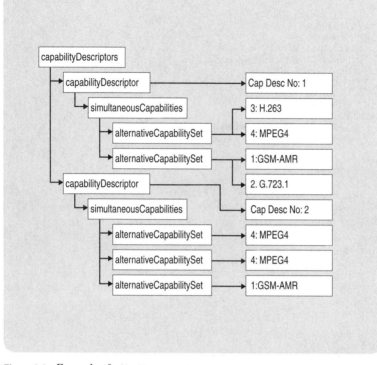

Figure 6.4 Example of `simultaneousCapabilities`.

When specifying capability table entries, if the receive video capability includes support for an option such as unrestricted motion vectors in the case of H.263, it must be able to accept and decode video encoded without this option.

An `alternativeCapabilitySet` is not restricted to containing references only to capability table entries of the same media type (e.g. all audio or all video). Nevertheless H.245 specifies that one or more capabilityDescriptors must exist that include an `alternative CapabilitySet` for each media type supported by the terminal, containing a `capabilityTableEntryNumber` for each capability of that media type. In the example shown in Figures 6.4 and 6.5 this is satisfied by `capabilityDescriptorNumber 1`.

H.245 suggests that the numbering of the `capabilityDescriptors` can be used to indicate the preference of the terminal: the lower the value of the `capabilityDescriptorNumber`, the more preferred it is. This can be used to do such things as signal that the terminal prefers H.263 at SQCIF (Sub-QCIF 128 × 96-pixel picture format) resolution to

```
Request terminalCapabilitySet {
    ...
    multiplexCapability = h223Capability{
     ...
    }
    CapabilityTable = 4 entries {
        [0] = {
          capabilityTableEntryNumber = 1
          capability = receiveAudioCapability genericAudioCapability {
          capabilityIdentifier = standard 0.0.8.245.1.1.1 -- GSM AMR
             ...
          }
        }
        [1] = {
          capabilityTableEntryNumber = 2
          capability = receiveAudioCapability g7231 {
           ...
          }
        }
        [2] = {
          capabilityTableEntryNumber = 3
          capability = receiveVideoCapability h263VideoCapability {
             ...
          }
        }
        [3] = {
          capabilityTableEntryNumber = 4
          capability = receiveVideoCapability genericVideoCapability {
          capabilityIdentifier = 0.0.8.245.1.0.0 -- MPEG4
             ...
          }
        }
    }
    CapabilityDescriptor = 2 entries {
        [0] = {
          capabilityDescriptorNumber = 1
          simultaneousCapabilities = 2 entries {
              [0] = 2 entries{
                  [0] = 1   -- GSM-AMR
                  [1] = 2   -- G.723.1
              }
              [1] = 2 entries{
                  [0] = 3   -- H.263
                  [1] = 4   -- MPEG4
              }
          }
        }
        [1]={
          capabilityDescriptorNumber = 2
          simultaneousCapabilities = 3 entries {
              [0] = 1 entries{
                  [0] = 1   -- GSM-AMR
              }
              [1] = 1 entries{
                  [0] = 4   -- MPEG4
              }
              [2] = 1 entries{
                  [0] = 4   -- MPEG4
              }
          }
        }
    }
}
```

Figure 6.5 Partial ASN.1 listing of terminalCapabilitySet request message.

H.263 at QCIF (Quarter Common Intermediate Format, 176 × 144 pixels) resolution by declaring a separate `capabilityTableEntry` for each of them, and referencing H.263 with SQCIF using a lower `capability- DescriptorNumber`.

6.3   Procedures and Signaling Entities

We have already seen in Chapter 4 that H.245 is based on the concept of *signaling entities* (SEs) dedicated to each of the procedures that H.245 supports. Under the supervision of an SE user process the SEs communicate with peer SEs in the remote terminal using H.245 messages to execute the procedure. Where the procedure can be independently executed in each of the two directions of the multimedia session, there are independent outgoing and incoming SEs. The SE user controls and initiates SEs, and they provide information on the outcome of the procedure to the SE user via the exchange of primitives.

There are 10 procedures defined in H.245. These procedures and their associated SEs are listed in Table 6.1. All the procedures are mandatory in H.324, H.324M, and 3G-324M except the Maintenance Loop procedure, which is optional. Some procedures can be executed multiple times in the course of a multimedia session. Examples of this are opening and closing logical channels and exchanging MTEs.

In Table 6.1 the SEs are classified as either persistent or not. For an SE that is not persistent, once the procedure has been executed the SE becomes idle. A persistent SE remains active. For example once capabilities have been exchanged the incoming and outgoing CESEs return to an idle state; however, when a unidirectional logical channel has been opened the outgoing SE (and the incoming SE in the peer terminal) remain in an active state until the logical channel is closed.

6.3.1   Signaling entities

To make sure that peer SEs keep in step with one another, request and response messages are generally used for SE–SE communication in preference to commands or indications, although a small number of commands and indications are used within the SEs. As discussed in Chapter 4, a request message requires that some action is taken by the peer SE and that the peer SE sends back a response message within a certain time. There is always a timer associated with the request/response interaction, which is started by the SE that sends the request. The response message takes the form of either an acknowledgment message indicating that the request was successful, or a reject message indicating that the SE has failed or refused to take the requested action. If the sending SE gets no response before the timer expires it will take some defined action.

Where a procedure can be attempted more than once in the event of failure, the SE may also include a counter that is incremented each time a procedure is initiated and compared with a maximum value before the procedure is reattempted.

Where the procedure has outgoing and incoming SEs, request messages are only sent from the outgoing SE, which also contains the timer and any counter (the incoming B-LCSE is the only incoming SE to include a timer, as we shall see). The incoming SE only ever sends response messages; where any commands and indications are used within a procedure, these are sent by the outgoing SE.

6.3.2 The SE user

The H.245 recommendation refers to the user of the MSDSE as the MSDSE user of the outgoing and incoming CESEs as the CESE user and so on. We will, however, consider there to be one user (the SE user), which is the supervisory process that communicates via primitives with all the SEs in H.245. This makes sense because some SEs are loosely coupled via the SE user. Examples include the close logical channel procedure, and the request multiplex entry procedures. The close logical channel procedure, which uses the outgoing and incoming CLSEs, causes the SE user to initiate a CloseLogicalChannel request using the outgoing LCSE or B-LCSE. In a similar way the request multiplex entry procedure, which uses the outgoing and incoming RMESEs, causes the SE to initiate a MultiplexEntrySend request using the outgoing MTSE.

The SE user process is not defined within H.245 or H.324. To begin establishing a session it must be informed or started at the request of the user of the terminal, when a bearer has been established, so it needs to have some direct or indirect interaction with the call establishment process and the user interface. The SE user must then coordinate the sequence of procedures needed to establish the session and communicate the results of the H.245 procedures to other components of the terminal such as the multiplexer and the codecs. For example, the outcome of multiplex entry exchange using the outgoing and incoming MTSEs must be communicated to the multiplexer via the SE user, to allow it to successfully demultiplex incoming MUX-PDUs.

H.245 defines **ERROR.indication** primitives within some of its procedures that are described as going to a management entity. For our purposes we assume that the SE user also receives and handles any error indications.

The SE user is the repository of information regarding the multimedia terminal, including such things as the codecs it supports, the multiplexer capabilities including mobile levels supported, ways in which the payload can be constructed and so on. This information is not stored in the SEs but is passed to them by the primitive that initiates the relevant procedure.

Many commands and indications are not associated with any SE, they have no need to have an associated state machine. A miscellaneous commands and indications process can be used to generate these PER encoded ASN.1 messages, when an appropriate primitive is generated by the SE user.

6.4   Description of Procedures

The procedures and associated SEs listed in Table 6.1 can be described in a number of ways. An outline of the operation of each procedure is provided here, concentrating on the behavior of the SEs when the procedure is successful. A complete functional description of the operation of each of the signaling entities involved is given in Appendix B, in a tabular form similar to that used in the hypothetical SE example of Chapter 2 and shown in Figures 2.8 and 2.9. The behavior of the SEs under all possible conditions, including timer expiry, receiving inappropriate messages, and the like is provided in Appendix B: the description of the procedures given here should be read in conjunction with Appendix B.

SDL descriptions of each of the procedures of H.245 are provided within the recommendation document. These cover many pages and are not included in this book. Interested readers are referred to the H.245 recommendation itself, the description of SDL diagrams in Chapter 2 should help in the interpretation of these diagrams. Alternatively the simple exercise of deriving the SDL diagrams from the descriptions in Appendix B may help in understanding the procedures.

6.4.1   Capability exchange

The capability exchange procedure is the first procedure to be initiated when establishing a 3G-324M session. Any subsequent procedures initiated by a terminal, such as opening a logical channel for video or audio, must respect the capabilities declared by the other terminal.

Capability exchange uses the outgoing and incoming *capability exchange SEs* (CESEs), and can be performed independently in each direction. It is initiated when the SE user instructs the outgoing CESE using the **TRANSFER.request** primitive. This passes parameters **PROTOID, MUXCAP, CAPTABLE, CAPDESCRIPTORS** to the outgoing CESE, allowing it to construct and send a `TerminalCapabitySet` request, containing `sequenceNumber`, a number to identify the request allowing responses to be correctly associated with it, `protocolID` specifying the version of H.245 in use, and `multiplexCapability`, `capabilityTable`, and `capabilityDescriptors`. When the

`TerminalCapabitySet` request is sent, the outgoing SE sets a timer T101 and waits for a response.

On receiving the `TerminalCapabitySet` request, the peer incoming CESE will inform its SE user, using the primitive **TRANSFER.indication(PROTOID, MUXCAP, CAPTABLE, CAPDESCRIPTORS)** to communicate the capabilities it has received. If its SE user accepts these capabilities, it indicates this using the primitive **TRANSFER.response**. This causes the incoming CESE to generate and send a `TerminalCapabilitySetAck` response containing a `sequenceNumber` that matches the `sequenceNumber` of the received request. If the outgoing CESE receives this before timer T101 expires and successfully checks the `sequenceNumber`, it uses the **TRANSFER.confirm** primitive to inform its SE user. The capability exchange procedure is then complete in this direction and the remote terminal is aware of the capabilities of the local terminal. Capability exchange in the other direction is a completely independent but identical process initiated by the remote SE user using the outgoing CESE of the remote terminal and the incoming CESE of the local terminal.

H.324 recommends that if the capability exchange procedure fails, it should be repeated at least two additional times before the attempt to establish a multimedia session is abandoned.

6.4.2 Master/slave determination

The master/slave determination procedure is initiated early in the session establishment process. H.324 states that the terminal capability exchange request must be the first message to be sent following synchronization of the multiplexers; however, it also specifies that the `MasterSlaveDetermination` request must be sent "at this time." Therefore the SE user need not wait for capability exchange to complete before initiating the master/slave determination procedure, though to avoid potential problems it makes sense for the SE user to wait until it is informed that a `TerminalCapabilitySet` request has been received from the remote terminal. The purpose of master/slave determination is to assign the status of master to one of the terminals, and the status of slave to the other, allowing the master to determine how any conflicts that do arise are resolved.

Because master slave determination cannot be performed independently in the two directions of the session, it cannot be split into outgoing and incoming SEs. Each terminal has a single *master/slave determination SE* (MSDSE).

The procedure for resolving which terminal is master and which is slave depends on two parameters. The first of these is the `terminalType`, the second is the `statusDeterminationNumber` (a random number generated by the terminal). If a 3G-324M user terminal is communicating with some 3G-324M network equipment instead of with another user terminal, there may be circumstances in which it is beneficial to have a predefined outcome to the master/slave determination process. The terminalType is used for this purpose. It has a value in the range 0 to 255, and is defined to be 128 for an end user terminal. The terminal with the highest value will always be set to master as a result of the master/slave determination process. Therefore by giving network equipment a lower value of terminalType than 128, it will always be slave; if its terminalType value is greater than 128, it will always be master when communicating with a user terminal.

If two terminals have equal terminalType values, the random number generated by the terminals is used to determine status. Whichever terminal has the higher random number becomes master.

The MSD procedure is initiated when the SE user instructs the MSDSE via the **DETERMINATION.request** primitive. This causes the MSDSE to send a MasterSlaveDetermination request message containing its terminalType and statusDeterminationNumber to the peer MSDSE, set a timer T106 and await a response. If the peer MSDSE is in the idle state when it receives the request, it will compare the received terminalType and statusDeterminationNumber with its locally stored values of **sv_TT** and **sv_SDNUM** to determine the status. It sends the response MasterSlaveDeterminationAck, containing the decision, sets its timer T106, and informs its SE user of the provisional result with the **DETERMINATION.indication(TYPE)** primitive. The decision in the MasterSlaveDeterminationAck represents the status of the receiving terminal. If decision = slave, the terminal receiving this Ack is slave. If the local MSDSE receives the Ack before its timer expires, it informs the SE user of the status using the **DETERMINATION.con-firm(TYPE)** primitive and sends a MasterSlaveDeterminationAck containing the decision to the peer MSDSE. If the peer MSDSE receives this Ack before its timer expires, and the decision accords with its determination of the status, it informs the SE user of the status using the **DETERMINATION.confirm(TYPE)** primitive and the master/slave determination procedure is complete.

The process of master/slave determination is unlike the other H.245 processes in that it features a three-way handshake; a request generates an acknowledgment that is in its turn acknowledged. It is, perhaps surprisingly, a rather complicated procedure, because both terminals can initiate the MasterSlaveDetermination request message and there is a chance that they can initiate them simultaneously. The condition that both generate the same value of random number also has to be handled, so the design of the procedure allows for multiple attempts, up to N100. H.324 recommends that the value of N100 should be at least 3. It also recommends that if the master/slave determination procedure fails, it should be repeated at least two additional times before the attempt to establish a multimedia session is abandoned.

Once the SE user has established that capability exchange and master/slave determination have completed, it can initiate the next steps in the process of session establishment. This could be to commence opening logical channels, or to exchange multiplex table entries. As has been previously mentioned, both these procedures need to complete before media can flow between the terminals.

6.4.3   Opening and closing logical channels

Once the incoming CESE has completed capability exchange, the SE user knows the capabilities of the remote terminal, and can determine which logical channels it is able to open, in accordance with these capabilities. Once master/slave determination has completed, it is in a position to open a logical channel.

Logical channels can be unidirectional or bidirectional. A unidirectional logical channel only has a forward logical channel. A bidirectional logical channel has a forward and reverse logical channel. The procedure for opening either type of logical channel uses H.245 messages with the same syntax, but the optional parameters specifying the reverse logical channel are not present in the case of unidirectional logical channels.

Unidirectional channels are opened and closed using the outgoing and incoming *logical channel SEs* (LCSEs). There is an instance of these SEs for each unidirectional logical channel the terminal can open. This means that unidirectional logical channels need not be opened in sequence, but their establishment can overlap. The same is true for bidirectional logical channels, which are opened using the outgoing and incoming *bidirectional LCSEs* (B-LCSEs).

To close established logical channels the close logical channel procedure is used, with its associated outgoing and incoming *close logical channel SEs* (CLCSEs). There is an instance of these SEs for each unidirectional and bidirectional logical channel. The numbers associated with logical channels must be unique, regardless of whether they are unidirectional or bidirectional. Once the close logical channel procedure has been used to successfully establish that a channel can be closed, it informs the SE user. The SE user then initiates the closure of the logical channel via the appropriate LCSEs or B-LCSEs.

The existence of a separate close logical channel procedure may seem at first unnecessary or overcomplex. The reason for having this procedure is to allow a logical channel to be closed by either terminal, overcoming the limitation that only the terminal that opened a channel can send the CloseLogicalChannel request.

Unidirectional logical channel signaling A unidirectional logical channel is always opened by the terminal that will use the channel to send information. To initiate opening a unidirectional logical the SE user instructs the outgoing LCSE via an **ESTABLISH.request(FORWARD_PARAM)** primitive. The **FORWARD_PARAM** information is used by the outgoing LCSE to construct and send an `OpenLogicalChannel` request containing the `forwardLogicalChannelNumber` and the `forward LogicalChannelParameters`.

The `forwardLogicalChannelParameters` are composed of the `dataType` and the `multiplexParameters`. The `dataType` further expands to specify whether it is `videoData` or `AudioData` (or more generally other data types not shown in Appendix A). These are mapped to the `VideoCapability` and `AudioCapability` data structures defined in the `TerminalCapabilitySet`. It is the responsibility of the SE user to ensure that these are populated with parameters consistent with the declared capabilities of the terminal the media is being sent to.

The `multiplexParameters` are `h223LogicalChannelParameters`, which specify the `adaptationLayerType` to be used and a BOOLEAN `segmentableFlag`, to indicate whether the logical channel is segmentable or not. If AL3 is used the number of `control-FieldOctets` and the `sendBufferSize` are included.

When the outgoing LCSE sends the `OpenLogicalChannel` request, it sets a timer T103 and waits for a response. If the peer incoming LCSE is in the released state when it receives the request it will use the **ESTABLISH.indication(FORWARD_PARAM)** primitive to inform its SE user and pass the received forward parameters to it. If the peer SE user accepts the request, it informs its incoming LCSE using the primitive **ESTABLISH.response**. The incoming LCSE then generates an `OpenLogicalChannelAck` containing the `forwardLogicalChannelNumber`. If the outgoing LCSE receives the Ack before its timer expires, the logical channel is successfully established.

Bidirectional logical channel signaling The establishment of a bidirectional logical channel has some similarities to the establishment of a unidirectional logical channel. It can be considered to comprise two associated unidirectional channels.

To initiate opening a bidirectional logical the SE user instructs the outgoing B-LCSE via an **ESTABLISH.request(FORWARD_PARAM, REVERSE_PARAM)** primitive. The **FORWARD_PARAM** and **REVERSE_PARAM** information is used by the outgoing LCSE to construct and send an `OpenLogicalChannel` request containing the `forwardLogicalChannelNumber`, `forwardLogicalChannel Parameters`, and `reverseLogicalChannelParameters`.

The difference between this request and the unidirectional openLogicalChannel request is the inclusion of the reverse LogicalChannelParameters. These are identical in structure to the forwardLogicalChannelParameters: they are composed of the dataType and the multiplexParameters. No logical channel number is specified for the reverse logical channel, because the outgoing B-LCSE has no way of determining what this should be. It is assigned by the remote terminal later in the procedure.

When the outgoing B-LCSE sends the OpenLogicalChannel request, it sets a timer T103 and waits for a response. If the peer incoming B-LCSE is in the released state when it receives the request, it will use the **ESTABLISH.indication(FORWARD_PARAM, REVERSE_PARAM)** primitive to inform its SE user and pass the received parameters to it. If the peer SE user accepts the request, it informs its incoming LCSE using the **ESTABLISH.response-(REVERSE_DATA)** primitive. The incoming LCSE then uses the **REVERSE_DATA** to generate an OpenLogicalChannelAck containing the forwardLogicalChannel Number and reverseLogicalChannel Parameters. The reverseLogicalChannelParameters in the Ack contains the reverseLogicalChannelNumber that the remote terminal has assigned to the bidirectional logical channel and should not be confused with the reverseLogicalChannelParameters in the request message.

The remote incoming B-LCSE then sends the OpenLogical ChannelAck message to the outgoing B-LCSE and sets its timer T103. If the outgoing LCSE receives the Ack before its timer expires, it informs the SE user of the logical channel number assigned by the remote terminal to the reverse logical channel with the **ESTABLISH. confirm(REVERSE_DATA)** primitive, and sends an OpenLogical ChannelConfirm indication message, containing the forwardLogical ChannelNumber. From the perspective of the originating terminal, the logical channel is now established. If the peer incoming B-LCSE receives the OpenLogicalChannelConfirm before its timer expires, it uses the **ESTABLISH.confirm primitive** to inform its SE user that the bidirectional logical channel is successfully established.

This three-way process of outgoing request, incoming Ack, and outgoing indication is similar to the behavior of the MSDSE, which uses outgoing request, incoming Ack, outgoing Ack messages to establish terminal status, except that an indication is used rather than a final Ack. I can almost hear you asking why? I don't know—it seems to be an inappropriate use of an indication message, because an action is required as a result (confirming that the channel is established). On the other hand, generating an Ack in response to an Ack is no less unusual.

Close logical channel procedure If the procedure of opening a logical channel fails before the channel is fully established, because the timer expires in the outgoing LCSE or B-LCSE, it will send a CloseLogicalChannel request to terminate the procedure. This contains the forwardLogicalChannelNumber and a source parameter. Because in this circumstance the LCSE or B-LCSE itself has automatically generated the request, the source is set to LCSE.

Once a logical channel is established, the message that causes the channel to be closed can only be sent by the outgoing LCSE or B-LCSE that opened it. To do this the SE user instructs the LCSE or B-LCSE using the **RELEASE.request primitive**. This causes the LCSE or B-LCSE to send a CloseLogicalChannel request containing the forwardLogicalChannelNumber and a source parameter, and set its timer T103. As in this circumstance the SE user has generated the request, the source is set to user. The peer incoming LCSE or B-LCSE uses the **RELEASE.indication(USER)** primitive to inform its SE user, and sends a CloseLogicalChannelAck. If the outgoing LCSE or B-LCSE receives this before its timer has expired, it uses the **RELEASE.confirm** primitive to inform the local SE user and the channel is closed.

The close logical channel procedure provides a mechanism that allows either terminal to close a logical channel, overcoming the limitation that the channel can only be closed by the outgoing LCSE or B-LCSE.

To initiate a close logical channel procedure, the SE user of the terminal that wishes to close the logical channel instructs the outgoing CLCSE via a **CLOSE.request** primitive. This causes the outgoing CLCSE to send a RequestChannelClose request containing the forwardLogicalChannelNumber and set its timer, T108. If the peer incoming CLSE is in the idle state when it receives the request, it will use the **CLOSE.indication** primitive to inform its SE user. If the peer SE user accepts the request, it informs its incoming CLCSE using the primitive **CLOSE.response**. The incoming CLCSE then sends a RequestChannelCloseAck containing the forwardLogicalChannel Number.

If the outgoing CLCSE receives the Ack before its timer expires, it informs the SE user using the **CLOSE.confirm** message and the procedure has completed. It is then the responsibility of the SE user that opened the logical channel to use the outgoing LCSE or B-LCSE to close it.

Resolving logical channel conflicts Conflicts can arise when opening logical channels. This can happen if both terminals make simultaneous requests to open bidirectional logical channels and the terminals only have the resources to support one bidirectional channel. Another

potential cause is that terminals may simultaneously try to open unidirectional logical channels that taken together cause a conflict with
the overall terminal capabilities. These conflicts are resolved by
making use of the results of master/slave determination. To understand
how these situations can be resolved we will look at a specific example of each.

In the first example we assume that we have two terminals that are
each capable of supporting a single video logical channel, in either H.263
or MPEG4 format, using AL3 (and therefore requiring a bidirectional
logical channel). Each terminal makes a simultaneous (or overlapping)
request to open a bidirectional logical channel with the forward and
reverse channels specified to contain video of either format. Both
requests cannot be successful because that would result in two video
streams in both directions. Clearly the intention is that there should be
just one bidirectional video channel, but both terminals have tried to
open it. In such circumstances the master terminal must reject the
request from the slave by sending `OpenLogicalChannelReject`. The
slave is able to work out for itself that there is a conflict, so if it does not
receive a rejection from the master terminal it should not reject the
request from the master but instead close the channel it has requested
by sending a `CloseLogicalChannel` request.

In the second example we assume that the master terminal supports
H.263 or MPEG4 format using AL2 (and therefore requiring a unidirectional logical channel). Its capabilities indicate that it is capable of,
in decreasing order of preference, receiving MPEG4 and H.263, but if
H.263 is to be received it must also be the format used for transmission. The slave terminal declares receive capabilities, in decreasing
order of preference, for H.263 and MPEG4. The master terminal makes
a request to open a unidirectional logical channel for H.263 and simultaneously the slave terminal makes a request to open a unidirectional
logical channel for MPEG4. These cannot both be successful due to the
receive-and-transmit limitation on H.263 in the master terminal. In
this case too, the master terminal must reject the request from the
slave by sending `OpenLogicalChannelReject`. The slave terminal
must then request a unidirectional logical channel for H.263 if a two
way conversational multimedia session is to be successfully established.

The conflict in this second example will not arise if the terminals
follow the recommendation in H.245 that the slave terminal should
attempt to open a logical channel for the master terminal's most preferred capability, and that the master should attempt to open a logical
channel for its own most preferred capability for which the slave has
declared support. This will result in both terminals requesting to open
H.263 from the outset.

6.4.4   Multiplex table entry procedures

The Multiplex table entries need to be communicated to allow each terminal to obtain the MTEs of the other and so be in a position to demultiplex received MUX-PDUs. There are two procedures associated with the transfer of MTEs: multiplex entry exchange and request multiplex entry. Multiplex entry exchange uses the outgoing and incoming Multiplex Table Entry SEs (MTSEs) to proactively send MTEs to the remote terminal. Request multiplex entry uses the Request Multiplex Entry SEs (RMESEs) to request MTEs from a remote terminal.

Table 6.1 shows that for these procedures there is an instance of the associated signaling entities for each of the up to 15 possible MTEs a terminal can support; however, the syntax of the associated H.245 requests and responses for both of these procedures defines message structures for both procedures that contain multiple MTEs. How this is achieved is left up to the implementer. One way of visualizing how this might be done for the MTSEs is shown in Figure 6.6, which shows a *message builder* function in front of the individual outgoing MTSEs. This takes the individual MTSE requests and forms them into a composite request. A similar function in front of the incoming MTSEs takes the composite request and breaks it into individual requests that are forwarded to the individual MTSEs. It also takes the responses from the incoming MTSEs and forms them into a composite response. When this

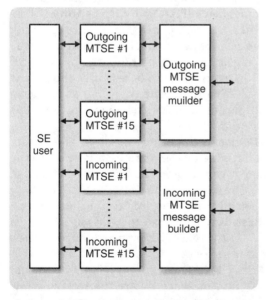

Figure 6.6   Forming single messages from multiple MTSEs.

is received at the outgoing side, it is decomposed into individual responses and forwarded to the individual outgoing MTSES. A similar model can be adopted for the RMESEs.

The relationship of the multiplex table entry send procedure to the request multiplex entry procedure is similar to the relationship between the open logical channel procedures and the close logical channel procedure. No multiplex table entries are transferred within the request multiplex entry procedure. If it completes successfully the requested terminal will then send the requested multiplex entries using the multiplex table entry send procedure.

Multiplex table entry exchange   Multiplex table entry exchange is performed independently in each direction. It is initiated when the SE user instructs the outgoing MTESE using the **TRANSFER.request(MUX-DESCRIPTOR)** primitive. This passes **MUX-DESCRIPTOR** to the outgoing MTESE, allowing it to construct a MultiplexEntrySend request, containing sequenceNumber, a number to identify the request allowing responses to be correctly associated with it, and a set of up to 15 MultiplexEntryDescriptors, one for each MTE. When the outgoing MTESE sends the request, it sets its timer T104 and waits for a response.

On receiving the MultiplexEntrySend request, the peer incoming MTESE will inform its SE user, using the primitive **TRANSFER.indication(MUX-DESCRIPTOR)** to communicate the capabilities it has received. If its SE user accepts these capabilities, it informs the peer incoming MTESE using the primitive **TRANSFER.response**. This causes it to generate and send a TerminalCapabilitySetAck response containing a sequenceNumber that matches the sequenceNumber of the received request. If the outgoing MTESE receives this before timer T104 expires, and successfully checks the sequenceNumber, it uses the **TRANSFER.confirm** primitive to inform the local SE user that the multiplex table entry exchange procedure has completed.

Multiplex table entry exchange in the other direction is an independent but identical process initiated by the remote SE user using the outgoing MTESE of the remote terminal and the incoming MTESE of the local terminal.

A terminal need not send all its MTEs in a single MultiplexEntrySend request. It can deactivate MTEs by sending a request with no elementList for the MultiplexTableEntryNumber within the MultiplexEntryDescriptor.

Rejection of a MultiplexEntrySend is not simply a binary decision. The MultiplexEntrySendReject message allows the receiving terminal to provide a set of MultiplexEntryRejectionDescriptions, allowing it to accept or reject each MTE independently.

Request multiplex entry   The `MultiplexEntrySend` request provides a way for a transmitting terminal to inform a receiving terminal of its MTEs. If the receiving terminal has a need to reobtain one or more MTEs, it will use the request multiplex entry procedure and the outgoing and incoming RMESEs.

Requests for MTEs to be re-transmitted are initiated when the SE user of the terminal that requires them instructs the outgoing RMESE using the **SEND.request** primitive, causing the outgoing RMESE to send a `RequestMultiplexEntry` request, containing a set of up to 15 `MultiplexTableEntryNumbers`, one for each MTE that it wants to receive. When the outgoing RMESE sends the request, it sets its timer T107 and waits for a response.

On receiving the `RequestMultiplexEntry` request, the peer incoming RMESE will inform its SE user, using the primitive **SEND.indication**. If its SE user accepts the request it informs the peer incoming RMESE using the primitive **SEND.response**. This causes it to generate and send a `RequestMultiplexEntryAck` response containing the relevant `MultiplexTableEntryNumbers`. If the outgoing RMESE receives this before timer T107 expires, it uses the **SEND.confirm** primitive to inform the local SE user.

Once the request multiplex entry procedure has completed, it is the responsibility of the SE user of the terminal sending the `RequestMultiplexEntryAck` to initiate a `MultiplexEntrySend` request for the appropriate MTEs as soon as possible.

6.4.5   Mode request

If a terminal has indicated its transmit capabilities in the capability exchange phase of multimedia session establishment, the mode request procedure provides a way for the receiving terminal to request that a particular mode be used to transmit information to it. As an example, if the transmitting terminal supports both QCIF and SQCIF picture sizes, the receiving terminal may use the mode request procedure to ask for QCIF to be used. If the terminal has not declared any transmit capabilities (which it need not, provided that whatever it does transmit is within the remote terminals receive capabilities), then the mode request procedure is of no use and must not be used. In any case, a terminal can refuse a request to transmit in a particular mode, although it seems likely that if the request is made it is with the purpose of providing an improved user experience, so if possible the terminal should comply with the request.

Referring to the Syntax of the `RequestMode` request in Appendix A, it contains a `sequenceNumber` to identify the message, and a number of `ModeDescriptions`. Each `ModeDescription` is a set of

ModeElements, which in their turn specify the media being referred to and the options being requested, and may optionally include a logical channel number and the H.223 adaptation layer type.

RequestMode requests with ModeElements that specify the logical channel number make requests specific to that logical channel, and may be used to change the mode for that channel. Such requests can only be made when the logical channel is open. When the logical channel number is not specified, the RequestMode request must contain a mode element describing each of the media types that are currently being transmitted. If it doesn't, this is interpreted as a request to stop transmission of this media type.

The response to the RequestMode request may be that the transmitting terminal sends an Ack agreeing ether to change to the receivers preferred mode (willTransmitMostPreferredMode), or that it will make some changes (willTransmitLessPreferredMode), or it may reject the request. If the request mode procedure is successful, it is the responsibility of the transmitting terminal to implement the requested changes as soon as possible by use of the appropriate procedures and messages. This may include closing and opening logical channels.

The request mode procedure is initiated when the SE user uses the **TRANSFER.request(MODE-ELEMENT)** primitive to instruct the outgoing MRSE to construct and send a RequestMode request. When the outgoing MRSE sends the request, it sets its timer T109 and waits for a response.

When the peer incoming MRSE receives the request it uses the primitive **TRANSFER.indication(MODE-ELEMENT)** to inform its SE user. If the SE user accepts the request, it uses **TRANSFER. response(MODE-PREF)** to instruct the incoming MRSE to send a RequestModeAck.

If the outgoing MRSE receives this before its timer expires it will use the **TRANSFER.confirm(MODE-PREF)** primitive to inform its SE user.

6.4.6 Round trip delay

The round trip delay procedure provides a way for terminals to determine whether the remote terminal is still responding and allows them to measure the time taken between a request being sent and a response being received. This can be used to adjust the values of the timers in the H.245 procedures described here.

The procedure makes use of the *round trip delay SE* (RTDSE), and is initiated when the SE user instructs the RTDSE using the **TRANSFER.request** primitive. This causes the RTDSE to send a RoundTripDelayRequest message, containing a sequenceNumber to allow the request to be identified, to the peer RTDSE and sets its timer T105. When the peer RTDSE receives the request, it does not inform its SE user but immediately sends back a `RoundTripDelayResponse`. When this is received by the RTDSE that made the request, it stops its timer, calculates the elapsed time, and uses the **TRANSFER. confirm(DELAY)** primitive to inform its SE user of the calculated round trip time.

6.4.7 Maintenance loop

The maintenance loop procedure is optional in H.324 (and therefore 3G-324M) and is intended for testing purposes. It allows a terminal to request that the remote terminal loops back the information being sent to it. The syntax of the `MaintenanceLoopRequest` message allows one of three types of loopback to be requested: `systemLoop`, `mediaLoop`, and `logicalChannelLoop`. The `systemLoop` option is for further study, so no function is associated with this and it should not be used.

The other two options specify a logical channel number, which must be associated with a bidirectional logical channel. The `mediaLoop` option indicates that the remote terminal is requested to take the information in the forward logical channel specified in the request, decode it, reencode it, and send it back using the reverse logical channel. The `logicalChannelLoop` option requests the remote terminal to take the information in the forward logical channel specified in the request and directly send it back using the reverse logical channel. If this procedure is supported, it is most likely to be useful for investigating problems with video over AL3.

The procedure uses the outgoing and incoming Maintenance Loop SEs and is initiated when the SE user instructs the MLSE using the **LOOP.request(LOOP-TYPE)** primitive. This causes the outgoing MLSE to send a `MaintenanceLoopRequest` message to the peer incoming MLSE and set its timer T102. When the incoming MLSE receives the request, it uses the **LOOP.indication(LOOP-TYPE)** primitive to inform its SE user. If the SE user accepts the request, it uses the **LOOP.response** primitive to instruct the incoming MLSE to send back a `MaintenanceLoopAck`. If this is received by the outgoing MLSE before the timer expires, it uses the **LOOP.confirm** primitive to inform its SE user.

After this procedure has completed the terminal associated with the incoming MLSE must implement the looping requested. The maintenance loop can be turned off by the SE user of the terminal that originally sent the `MaintenanceLoopRequest`. It does this by using the **RELEASE.request** primitive to instruct the outgoing MLSE to send a `MaintenanceLoopOffCommand`. When the peer incoming MLSE receives this command it uses the **RELEASE.indication** primitive to inform the SE user, which must then take the necessary action to ensure the terminal configuration is changed to switch looping off.

6.5   H.245 Commands and Indications

H.245 has some commands and indications that are not associated with any SEs or procedures but are important for the operation of a 3G-324M terminal. The implementation of these commands is not described within the recommendation. Their use is described here.

6.5.1   Flow control

The `FlowControlCommand` allows a terminal to command the remote terminal to limit the bit rate of the information it is sending. This command can apply to a specific logical channel or to the whole of the multiplex.

If the command applies to a single logical channel, other logical channels are free, where appropriate, to increase their bit rate to occupy the bandwidth freed up. The command is most likely to be applied to a logical channel for video. Support for the command is mandatory in H.324.

When a terminal receives a `FlowControlCommand`, it must respond with a `FlowControlIndication` to indicate its new settings.

6.5.2   End session

3G-324M calls are ended using the `EndSessionCommand`. Within the command H.245 specifies a choice of a number of options for actions the terminal should subsequently take or modes it should enter. Appendix A shows that for 3G-324M only one of these options remains—`disconnect`. Once a terminal has issued this command, it must send no further H.245 messages, but terminate the session, and the call.

6.5.3   Miscellaneous command

The `MiscellaneousCommand` is used for a number of purposes, which are indicated by the choice of option within it. These include `videoFastUpdatePicture` and other options relating to partial picture updating. These options can be used to obtain Intra coded video information when a receiving terminal believes that its video decoder is suffering from corrupt or lost data. The 3GPP implementers guide TR 26.911 recommends that H.263 encoders in 3G-324M terminals should respond to all video update options that are specified within this command.

Another video related miscellaneous command option is `video TemporalSpatialTradeOff`, which allows terminals to modify the trade off between bits per frame (spatial resolution) and frames per second (temporal resolution) for received video. There is an integer parameter in the range 0 to 31 associated with it, where 0 corresponds to the highest spatial resolution and 31 corresponds to the highest frame rate.

The miscellaneous command option maxH223MUXPDUSize can be used to limit the size of received MUX-PDUs to the number of octets specified by the associated integer parameter.

6.5.4 H.223 multiplex configuration

We have seen in Chapter 5 that the multiplexer has a number of optional modes of operation. It can operate at different mobile levels and can use optional double flags at ML1 and an optional three octet header at ML2.

The procedure for synchronizing the multiplexers of two communicating terminals at the highest common mobile level was described in Chapter 4. To avoid ambiguity at the start of the session no optional modes can be used. The multiplexer options can be changed once the session has been established, and mobile levels can be switched using the H223MultiplexReconfiguration command.

The command can only be used if the remote terminal has declared support for changing multiplexer mode through the modeChange Capability parameter within mobileOperationTransmit Capability, within H223Capability in its Terminal-CapabilitySet.

The procedure for changing mobile level settings is described in Annex C of H.324. The H223MultiplexReconfiguration command is sent specifying the mobile level or option to change to. The receiving terminal must stop transmitting any MUX-PDUs and instead send the ones complemented stuffing sequence for the current level, at least ten times and for no longer than 500 ms, followed by switching to the new level. The receiving terminal detects the transition from ones complemented stuffing sequence for the old level to opening flag for the new level to resynchronize. The terminal changing level should not itself issue an H223MultiplexReconfiguration command whilst this procedure is completing.

This procedure implies that different mobile levels can be used in each direction. Mobile levels can be changed to higher or lower levels provided that they remain within the declared capability of the transmitting terminal.

6.5.5 Function not supported indication

The FunctionNotSupported indication is used when an H.245 message is received which is not recognized. The indication includes a cause, which may be one of syntaxError, semanticError or unknown Function, and the option to return the function in PER format.

An example might be that if a terminal receives a Maintenance LoopRequest, which is an optional procedure, and it does not support that function, it may send a FunctionNotSupported indication, with the cause set to unknownFunction.

6.5.6 Miscellaneous indication

The MiscellaneousIndication message can be used to indicate the current videoSpatialTradeOff setting for a H.263 video logical channel. If a terminal has declared TemporalSpatialTradeOffCapability, it must send this indication message when a logical channel for H.263 is opened and every time it makes an adjustment. 3GPP recommends that MPEG4 encoders should also respond to videoSpatialTradeOff within a MiscellaneousCommand. Support for TemporalSpatialTradeOff Capability for MPEG4 cannot be explicitly declared in the capabilities; 3GPP recommends that MPEG4 encoders support it by default.

If H.263 with Annex I as recommended by 3GPP is being used, videoNotDecodedMBs can be used within a Miscellaneous Indication message to indicate that some Macroblocks with transmission errors have been received. The encoder may be able to use this information to improve the coded H.263 information it subsequently sends.

6.5.7   H.223 logical channel skew

The H223SkewIndication message allows a terminal to communicate the relative delay in milliseconds between two logical channels. Typically these will be an audio and a video channel. The indication allows a terminal that includes an optional delay in its audio receive path to adjust the value of this delay to achieve better lip synchronization.

6.5.8   Vendor identification

The VendorIndication message may be used by a terminal manufacturer (or the manufacturer or developer of the H.245 component within the terminal) to identify the manufacturer, the product, and the version number.

6.5.9   User input indication

The UserInputIndication message allows alphanumeric characters to be sent by a terminal, providing the terminal with a capability analogous to the DTMF capability of GSTN phones. It can be used to transfer user choices for applications such as video mail. H.324 specifies that support for the transmission of the characters 0–9, *, and # (the characters on a handset keypad) is mandatory.

This indication can either transfer a userInputSupport Indication consisting of a string of one or more characters or alternatively it can provide a closer representation of a DTMF key press by using the signal structure, with signalType being a single character and duration indicating the length of time the key is held down.

UserInputIndication messages can only be sent if support for this capability has been declared in the received TerminalCapabilitySet of the remote terminal.

6.6   H.245 Generic Capabilities

The syntax of the H.245 recommendation provides explicit codepoints for the H.263 video codec and for the G.723.1 audio codec; however, it does not do this for the other codecs in 3G-324M terminals—the GSM-AMR audio codec and the MPEG4 video codec. To signal information relating to these codecs, H.245 has to make use of genericVideoCapabilities and genericAudioCapabilities. Both of these are defined using the GenericCapability data structure shown in Appendix A. This is used in conjunction with Appendices of the H.245 recommendation to build a capability for codecs that do not have explicit codepoints. H.245 Appendix E "ISO/IEC 14496-2 Capability Definitions" provides the required information for MPEG4 video, and Appendix I "GSM Adaptive Multirate Capability Definitions" provides the information for GSM-AMR.

The appendices of H.245 provide the CapabilityIdentifier for the relevant standard, specify which of the fields in the Generic Capability are required, and define the set of GenericParameters that the codec uses.

Generic capabilities are referenced within the Terminal CapabilitySet message, the LogicalChannelMessage and the RequestMode message. Each GenericParameter may be present in some of these messages but not others; this is specified in the definition of each GenericParameter in the appropriate appendix of the H.245 recommendation. Each GenericParameter is classified as collapsing or nonCollapsing: this relates to how it is handled in multipoint applications. It needs to be specified in the message but otherwise can be ignored for our purposes.

For GSM-AMR the GenericParameters correspond to such things as the maximum number of GSM-AMR frames per AL-SDU (parameterIdentifier 0), the bit rate (parameterIdentifier 1, with parameterValue set to a value from 0 to 7, where 0 represents the lowest 4.75-kbit/s mode and 7 represents the highest 12.2-kbit/s mode), and several other optional GenericParameters relating to the type of comfort noise.

For MPEG4 the GenericParameters include the Profile and Level of the MPEG4 codec (parameterIdentifier 0, parameterValue 8, corresponding to Simple Profile Level 0), the set of tools to be used by the decoder (parameterIdentifier 1), and the decoder configuration information (parameterIdentifier 2). The last of these contains an octet string representing the start codes for the expected MPEG4 stream. This information will also be contained in the stream itself.

6.7 Chapter Summary

We have looked at the underlying syntax and meaning of H.245 messages and at the purpose and operation of the procedures defined in H.245 for establishing a multimedia session, referring extensively in this chapter to Appendices A and B of this book.

Appendix A provides a description of the semantics of the messages most likely to be encountered in a 3G-324M call. Appendix B provides a complete description of the procedures, including how they behave when the procedure does not proceed as smoothly as is described in this chapter.

Together with the preceding chapters this should have given you sufficient background to understand more easily what happens in the call example that is presented in Chapter 7.

Session Walkthrough

Having looked at the components that make up 3G-324M, the best way to understand how it works is to look at an example of multimedia session establishment. We will consider the case of a call between two identical mobile video phones that support mobile levels 0, 1, and 2; GSM-AMR and G.732.1 audio over AL2; and H.263 and MPEG4 video over AL3.

For the purposes of this example we will assume that we have a tool that can capture and display the H.245 messages, which flow between the terminals in human-readable ASN.1 format. The availability of tools for this purpose is discussed in Chapter 8.

7.1 Phases of the Call

The first step is the establishment of the bearer. This is outside the scope of 3G-324M, and has been discussed in Chapter 4. Once the bearer has been established correctly, the end user terminals will go through the following stages to set up the 3G-324M session:

1. Establishment of initial mobile level

2. Terminal capability exchange

3. Master/slave determination

4. Open logical channels

5. Multiplex table exchange

When these steps have been completed, media can start to flow between the terminals.

Establishing the initial mobile level has already been discussed in Chapter 4. The two terminals send out the stuffing sequence corresponding to the highest mobile level they support, and examine what is being sent to them. The terminals should detect at least five consecutive occurrences of the stuffing flag to ensure that it is detected reliably. In this example the terminals are identical and both send identical stuffing sequences corresponding to ML2.

We assume that mobile level detection has proceeded successfully and that the two terminals are transmitting and receiving at ML2. They can identify MUX-PDU boundaries in their incoming bit stream and demultiplex the received data. The only types of MUX-PDU that can be handled or understood at this stage are empty MUX-PDUs (8 octets with the value E1 4D 00 00 00 in hexadecimal) or MUX-PDUs formed using the inbuilt LCN0 and MTE0, to allow the exchange of H.245 control protocol messages.

Because we are considering the example of two identical mobile video phones, we will only look at the messages sent by the outgoing SEs of party A and the incoming SEs of party B. There will be an almost identical set of messages associated with the outgoing SEs of party B and the incoming SEs of party A. In the case where the two mobile video phones are not identical, similar messages will be sent, but the contents of the messages sent in each direction are very likely to be different.

In real life the procedures involved in session establishment operate in an asynchronous and overlapping way. This means that requests, responses and other messages associated with different procedures are likely to be intermingled and makes a listing of the H.245 messages slightly less easy to interpret in practice than in this idealized example.

To make it easier to understand what is going on, we have created an idealized record of the session establishment, in which the request messages receive immediate response messages and each procedure completes before the next procedure starts. In this chapter we will examine this record step by step.

Some comments have been added to the listings, preceded by two dashes to identify them, as follows:

```
-- This is a comment
```

7.1.1 Terminal capabilities exchange

The first step in establishing the multimedia session is for the terminals to exchange information on the capabilities that they have.

Capability exchange is initiated when the outgoing CESE generates a `terminalCapabilitySet` message and sends it to the remote terminal where it is processed by the peer incoming CESE. Since this is a

```
-- MESSAGE SENT BY OUTGOING CESE: TERM CAPS EXCHANGE (FIRST SECTION)

Request terminalCapabilitySet
{
    sequenceNumber = 1
    protocolIdentifier = 0.0.8.245.0.10
    multiplexCapability = h223Capability
    {
        transportWithI-frames = FALSE
        videoWithAL1 = FALSE
        videoWithAL2 = FALSE
        videoWithAL3 = TRUE
        audioWithAL1 = FALSE
        audioWithAL2 = TRUE
        audioWithAL3 = FALSE
        maximumAl2SDUSize = 128 -- allows up to about 4 audio frames
        maximumAl3SDUSize = 4096 - used for video
        h223MultiplexTableCapability
        {
        Basic
        }
        maxMUXPDUSizeCapability = TRUE
        nsrpSupport = TRUE                -- mandatory for 3G-324M
        mobileOperationTransmitCapability
        {
        modeChangeCapability = TRUE
        h223AnnexA = TRUE          -- mandatory for 3G-324M
        h223AnnexADoubleFlag = FALSE
        h223AnnexB = TRUE          -- mandatory for 3G-324M
        h223AnnexBwithHeader = FALSE
        }
    bitRate = 640 -- 64 kbit/s
}
```

Figure 7.1 First part of terminal capabilities exchange request.

large message we will consider it in three sections. The first of these is shown in Figure 7.1.

The `sequenceNumber` increments each time a `Terminal CapabilitySet` message is sent during the session. It is used in responses to the message, enabling them to be identified as responses to this message. The `protocolIdentifier` defines the protocol in use. The meaning of the fields is ((0: itu-t). (0: recommendation). (8: h-series). (0: h245).(10: version). Since the protocol in use will always be one or other version of H.245 only the last field of this identifier may change from phone to phone.

Because all 3G-324M compliant terminals use H.223 as the multiplexing protocol, `multiplexCapability = h223Capability` will always be true, though the features of a particular H.223 implementation will vary. These are defined in the remainder of this first section of the `TerminalCapabilitySet` message. The `transportWithI-frames` line has nothing to do with video coding. It is always set to FALSE because this indicates that SRP or NSRP is being used. The lines following this indicate which adaptation layers support which media type, and set maximum sizes of AL-SDUs that the multiplexer is capable of receiving. In this case video over AL3 and audio over AL2 are supported. The maximum AL-SDU sizes are 128 octets for AL2 and 4096 octets for AL3.

The capability of the multiplexer to support basic or enhanced MTEs is signaled by h223MultiplexTableCapability, which in this case specifies support only for basic MTEs. Use of NSRP, which is mandatory for mobile levels other than ML0, is indicated by nsrpSupport being set to TRUE.

The mobileOperationTransmitCapability defines the mobile levels that the multiplexer supports and, via modeChangeCapability, whether it is possible to change levels or options within levels using the H223MultiplexReconfiguration command described in Chapter 6. For a terminal to be 3G-324M compliant it must support mobile levels 0, 1, and 2, though support for options such as h223AnnexADoubleFlag and h223AnnexBwithHeader within these levels is not mandatory.

The bitRate value is expressed in multiples of 100 bit/s, so 640 means that the overall bit rate of the multiplexer is limited to 64 kbit/s, matching the bandwidth of available bearer.

The second section of the TerminalCapabilitySet message, defining the set of capabilities that the terminal supports, is shown in Figure 7.2.

The CapabilityTable has a number of entries, one for each capability that is supported. Each elemental capability is given a number, assigned by capabilityTableEntryNumber. The capability structure provides details of the capability including the support for any associated parameters.

In our example, the first capability declared is the capability to receive GSM-AMR, a mandatory capability for any 3G-324M compliant terminal. This is defined using the genericAudioCapability structure with a single generic parameter as specified in Annex I of the H.245 recommendation indicating that there is one GSM-AMR frame in each AL-SDU.

The second capability listed is to receive G.723.1, an optional codec for 3G-324M, with one frame per AL-SDU and silence suppression enabled.

The third capability declares the ability to receive H.263. It shows that the decoder can handle QCIF format at about 10 frames per second (a frame interval of $3 \times (1/29.7)$ s) or SQCIF format at about 15 frames per second. It can accept video at a maximum bit rate of 48 kbit/s, as specified by the integer value 480 assigned to maxBitRate, which is in multiples of 100 bit/s. It supports spatial-temporal trade-off but not arithmetic coding, unrestricted motion vectors, advanced prediction, pbframes, or error compensation.

The fourth capability supported is MPEG4 video (ISO/IEC 14496-2) via genericVideoCapability, with one generic parameter, as specified in

```
-- MESSAGE SENT BY OUTGOING CESE: TERM CAPS EXCHANGE (SECOND SECTION)
    CapabilityTable = 5 entries
    {
        [0] =
        {
         capabilityTableEntryNumber = 1
         capability = receiveAudioCapability genericAudioCapability
         {
             capabilityIdentifier = standard 0.0.8.245.1.1.1 --GSM AMR
             maxBitRate = 122                              -- 12.2 kbit/s
             collapsing = 1 entries
             {
                 [0]=
                 {
                   parameterIdentifier = standard 0 -- frames per AL-SDU
                   parameterValue = unsignedMin 1  -- set to 1 per SDU
                 }
             }
         }
        }
        [1] =
        {
         capabilityTableEntryNumber = 2
         capability = receiveAudioCapability g7231
         {
             maxAl-sduAudioframes = 1
             silenceSuppression = TRUE
         }
        }
        [2] =
        {
         capabilityTableEntryNumber = 3
         capability = receiveVideoCapability h263VideoCapability
         {
             sqcifMPI = 2                          -- target 15 frames/s
             qcifMPI = 3                           -- target 10 frames/s
             maxBitRate = 480
             unrestrictedVector = FALSE
             arithmeticCoding = FALSE
             advancedPrediction = FALSE
             pbFrames = FALSE
             temporalSpatialTradeOffCApability = TRUE
             errorCompensation = FALSE
         }
        }
        [3] =
        {
         capabilityTableEntryNumber = 4
         capability = receiveVideoCapability genericVideoCapability
         {
             capabilityIdentifier = 0.0.8.245.1.0.0     -- MPEG4
             maxBitRate = 480                           -- 48 kbit/s
             noncollapsing - 1 entries
             {
                 [0]=
                 {
                   parameterIdentifier = standard 0 -- profileAndLevel
                   parameterValue = unsignedMax 8   - SimpleProfile@L0
                 }
             }
         }
        }
        [4] =
        {
         capabilityTableEntryNumber = 5
         capability =receiveAndTransmitUserInputCapability basicString
        }
    }
```

Figure 7.2   Second part of terminal capabilities exchange request.

Annex E of the H.245 Recommendation indicating that it supports simple profile level 0.

All these capabilities are receive capabilities. The terminal could also separately declare transmit capabilities but does not have to. If the terminal is limited to the same coding standard for both transmitted and received media then it would declare receiveAndTransmit AudioCapability instead of receiveAudioCapability and receiveAndTransmitVideoCapability instead of receiveVideo Capability. This may be the case for many video telephones, due to the limited processor and memory resources available.

The final capability declared is receiveAndTransmitUser InputCapability in the format basicString, allowing the terminal to send and receive the ASCII characters 0–9, *, and #. This provides the mobile video telephone with the capability to interact with services such as video mail or video streaming services, and select the video mail or clip to be played.

GSM-AMR is preferred over G.723.1, and H.263 is preferred over MPEG4 because they have been assigned a lower capability TableEntryNumber.

The final part of the TerminalCapabilitySet message is given in Figure 7.3.

```
-- MESSAGE SENT BY OUTGOING CESE: TERM CAPS EXCHANGE (LAST SECTION)

  CapabilityDescriptor = 1 entries
  {
     [0]=
     {
       capabilityDescriptorNumber = 1
       simultaneousCapabilities = 3 entries
       {
          [0] = 2 entries
          {
             [0] = 1     -- GSM-AMR
             [1] = 2     -- G.723.1
          }
          [1] = 2 entries
          {
             [0] = 3     -- H.263
             [1] = 4     -- MPEG4
          }
          [2] = 1 entries
          {
             [0] = 5     -- User Input Indication
          }
       }
     }
  }
} -- End of Request TerminalCapabilitySet Message
```

Figure 7.3   Final part of terminal capabilities request.

```
-- MESSAGE SENT BY PEER INCOMING CESE: TERM CAPS EXCHANGE SUCCESSFUL

Response TerminalCapabilitySetAck
{
       SequenceNumber = 1
}
```

Figure 7.4   Terminal capabilities exchange acknowledgement.

This contains a `CapabilityDescriptor` structure, consisting of a list of `simultaneousCapabilities` that the terminal can support at the same time. These are in their turn composed of a list of alternative capabilities, identified by their `capabilityTableEntryNumber`. It declares that the terminal is capable of simultaneously supporting {GSM-AMR or G.723.1}, {H.263 or MPEG4}, and {User Input Indication}.

When the outgoing CESE has sent this message, it sets timer T101 and waits for a response. If the terminal capabilities are successfully received and accepted by the remote terminal, the incoming CESE at the remote terminal will send a `TerminalCapabilitySetAck` response, as shown in Figure 7.4.

This contains the sequence number of the `TerminalCapabilitySet` request that it is responding to, so that the outgoing CESE knows which `TerminalCapabilitySet` request it is acknowledging. If a valid Ack is received the SE user is notified and the terminal can move on to the next stage of session establishment.

If the remote terminal does not accept the `TerminalCapabilitySet` request, its incoming CESE sends a `TerminalCapabilitySetReject` response, as shown in Figure 7.5.

This message includes both the sequence number of the `TerminalCapabilitySet` request that it is responding to and a cause, which is unspecified in this example. The `Terminal CapabilitySetReject` response message may give causes such as `undefinedTableEntryUsed`, `descriptorCapacityExceeded` or `tableEntryCapacityExceeded`, and in the last case provide further information. These causes may be seen in the case of implementation

```
-- MESSAGE SENT BY PEER INCOMING CESE: TERM CAPS EXCHANGE FAILURE
Response TerminalCapabilitySetReject
{
sequenceNumber = 1
cause = unspecified <<null>>
}
```

Figure 7.5   Terminal capabilities exchange rejection.

```
-- MESSAGE SENT BY OUTGOING CESE: NO RESPONSE BEFORE TIMER EXPIRY

Indication TerminalCapabilitySetRelease{}
```

Figure 7.6   Terminal capabilities exchange release indication.

errors or limitations in the ability of the remote handset to store the information sent.

If no response is received before the T101 timer in the outgoing CESE expires, the outgoing CESE sends a `TerminalCapabilitySetRelease` indication as shown in Figure 7.6.

Because `TerminalCapabilitySetRelease` is an indication rather than a request, there is no response to this message. After a response is received or a `TerminalCapabilitySetRelease` is sent, the outgoing CESE returns to the idle state. As mentioned in Chapter 6, it is recommended that if the CESE fails it should be reattempted twice. The action that is then taken is outside the scope of the standards and is left to the implementer. Possible further courses of action may be to try again with a more limited capability set, or to clear the call down.

7.1.2   Master/slave determination

Assuming that the terminal capability exchange has completed successfully, the next stage in setting up the session is to determine which terminal will be master and which terminal will be slave. To start master/slave determination one of the terminals, most likely the one that completes terminal capability exchange first, sends a `MasterSlaveDetermination` request and sets its timer T106. An example of this message is shown in Figure 7.7.

In the request shown above `terminalType` is set to the value of 128 indicating that it is a user terminal, so, because it is also set to this value on the remote mobile video telephone, the status `DeterminationNumber`, a random number 3 octets wide, is used to determine which terminal will be master. The terminal that receives the `MasterSlaveDetermination` request makes the decision on the status

```
-- MESSAGE SENT BY OUTGOING MSDSE: COMMENCE MASTER SLAVE DET

Request MasterSlaveDetermination
{
    terminalType = 128
    statusDeterminationNumber = 13120659
}
```

Figure 7.7   Master/slave determination request.

```
-- MESSAGE SENT BY PEER MSDSE: MASTER SLAVE DET SUCCESSFUL

Response MasterSlaveDeterminationAck
{
    decision = slave <<null>>    -- receiving terminal is slave
}
```

Figure 7.8 Master/slave determination acknowledgement (incoming).

of the terminals by comparing the value of status DeterminationNumber it receives with its own internal random number. If it can make a clear decision it sends the Master SlaveDeterminationAck response message of Figure 7.8.

The status indicated by decision, which is either master or slave, is the status of the terminal that receives this response message. In our example the remote terminal is the master. The determination process is not complete until the MSDSE that receives the first Ack replies with an Ack of its own, with decision set to the opposite of the status it received, as shown in Figure 7.9.

The terminal that receives this second Ack checks it for consistency; if so the master/slave determination step has completed successfully. Terminals must not initiate any procedures that rely on knowing the result of master/slave determination (for example opening a bidirectional logical channel) until they have received confirmation that the result is known and accepted by both terminals.

If the terminal receiving the MasterSlaveDetermination request is unable to determine status, it will send a MasterSlave DeterminationReject response as shown in Figure 7.10.

This occurs if the terminals have identical random numbers, which is indicated by the cause being set to identicalNumbers. Provided that it has not made more than N100 attempts, the terminal will repeat the attempt to establish master/slave status—generating a new random number.

If no response is received to the MasterSlaveDetermination message before timer T106 expires, it sends a MasterSlave DeterminationRelease as shown in Figure 7.11.

In the event of failure to determine which terminal is master and which terminal is slave the MSDSE should retry at least two further times

```
-- MESSAGE SENT BY MSDSE: MASTER SLAVE DET SUCCESSFUL

Response MasterSlaveDeterminationAck
{
    Decision = master <<null>>    -- receiving terminal is master
}
```

Figure 7.9 Master/slave determination acknowledgement (outgoing).

```
-- MESSAGE SENT BY PEER MSDSE: MASTER SLAVE DET UNSUCCESSFUL

Response MasterSlaveDeterminationReject
{
      Cause = identicalNumbers <<null>>
}
```

Figure 7.10   Master/slave determination rejection.

before abandoning the attempt and informing the SE user, which must then end the session by sending the command EndSessionCommand.

7.1.3   Open logical channel (audio)

Once master/slave determination has been completed, either logical channels can be opened or multiplex table entries can be exchanged.

In our example we will open logical channels first, starting with an audio logical channel. Multiple logical channels do not have to be opened sequentially, the process of opening them can overlap.

We will start by opening a unidirectional audio logical channel as LCN1. To do this the terminal chooses to request GSM-AMR at a maximum rate of 12.2 kbit/s with one frame per AL-PDU, using the AL2 adaptation layer of the multiplexer, without sequence numbers. The resulting OpenLogical Channel request message that is sent is shown in Figure 7.12.

In this message segmentableFlag = FALSE means that the AL-PDUs presented to the multiplex layer are not segmentable. After sending the request, the outgoing LCSE sets timer T103 and waits for a response.

The incoming LCSE of the terminal that receives this message can either accept the request and send an Ack as shown in Figure 7.13, or reject the request as shown in Figure 7.14.

If the request is rejected a number of causes may be given via the cause field in the OpenLogicalChannelReject structure. In this example the request is to open a unidirectional audio channel so unsuitableReverseParameters is not a relevant cause, and the OpenLogicalChannel request did not include a replacementFor field so replacementForRejected is also not relevant. If cause is set to dataTypeNotSupported, dataTypeNotAvailable, unknownData Type, or dataTypeALCombinationNotSupported, this points to an implementation problem where the SE user has failed to correctly inter- pret or use the terminal capabilities supplied by the remote terminal. The most likely cause in this example is either unspecified or masterSlaveConflict. If cause is set to masterSlaveConflict the reason is likely to be that simultaneous (or overlapping) requests to open an audio logical channel have been sent by both terminals, but with different AudioCapability values. This circumstance arises when the terminals are only capable of supporting the same AudioCapability

```
-- MESSAGE SENT BY MSDSE: NO RESPONSE BEFORE TIMER EXPIRY

Indication MasterSlaveDeterminationRelease{}
```

Figure 7.11   Master/slave determination release indication.

```
-- MESSAGE SENT BY OUTGOING LCSE: OPEN LOGICAL CHANNEL REQUEST

Request OpenLogicalChannel
{
    forwardLogicalChannelNumber = 1
    forwardLogicalChannelParameters =
    {
        dataType = audioData genericAudioCapability
        {
          capabilityIdentifier = standard 0.0.8.245.1.1.1   --GSM-AMR
          maxBitRate = 122
          collapsing = 1 entries
          {
             [0]=
             {
              parameterIdentifier = standard 0
              parameterValue = unsignedMin 1
             }
          }
        }
    }
        multiplexParameters = h223LogicalChannelParameters
        {
          adaptationLayerType = al2WithoutSequenceNumbers <<null>>
          segmentableFlag = FALSE
        }
    }
}
```

Figure 7.12   Open logical channel request.

```
-- MESSAGE SENT BY PEER INCOMING LCSE: OLC SUCCESSFUL

Response OpenLogicalChannelAck
{
    forwardLogicalChannelNumber = 1
}
```

Figure 7.13   Open logical channel acknowledgment.

```
-- MESSAGE SENT BY PEER INCOMING LCSE: OLC UNSUCCESSFUL

Response OpenLogicalChannelReject
{
    forwardLogicalChannelNumber = 1
    cause = unspecified <<null>>
}
```

Figure 7.14   Open logical channel rejection.

```
-- MESSAGE SENT BY OUTGOING LCSE: NO RESPONSE BEFORE TIMER EXPIRY

Request CloseLogicalChannel
{
    forwardLogicalChannelNumber = 1
    source = lcse <<null>>
}
```

Figure 7.15   Close logical channel request.

in both directions, as indicated by `receiveAndTransmit`
`AudioCapability` in the `CapabilityTableEntry` within the
`TerminalCapabilitySet`. If this happens the master terminal will
send an `OpenLogicalChannelReject` message with `cause` set to
`masterSlaveConflict` to the slave terminal in response to the
`OpenLogicalChannel` request from the slave.

If the outgoing LCSE receives no response before timer T103 expires,
it sends a `CloseLogicalChannel` request, as shown in Figure 7.15.

Possible `source` settings are `lcse` as in this example, where the outgoing LCSE has automatically generated the `CloseLogicalChannel`
request, and `user`, where the outgoing LCSE has been requested by the
SE user to close the logical channel. This is a request, so the peer incoming LCSE sends an Ack in response that has the format of Figure 7.16.

7.1.4   Open bidirectional logical channel (video)

In our example video is carried over the AL3 adaptation layer, which implicitly means it requires a bidirectional logical channel. The messages that flow
between the signaling entities are based on the same ASN.1 structure and
syntax definition as those used to open unidirectional logical channels,
except that they include `reverseLogicalChannelParameters`.

To open a bidirectional logical channel as LCN2 the requesting terminal sends an `OpenLogicalChannel` request, as shown in Figure 7.17.

```
-- MESSAGE SENT BY INCOMING LCSE: ACK FOR CLOSE CHANNEL REQUEST

Response CloseLogicalChannelAck
{
    forwardLogicalChannelNumber = 1
}
```

Figure 7.16   Close logical channel acknowledgment.

```
-- MESSAGE SENT BY OUTGOING B-LCSE: B-OLC REQUEST

Request OpenLogicalChannel
{
    forwardLogicalChannelNumber = 2
    forwardLogicalChannelParameters =
    {
        dataType = videoData h263VideoCapability
        {
          qcifMPI = 2
          unrestrictedVector = FALSE
          arithmeticCoding = FALSE
          advancedPrediction = FALSE
          pbFrames = FALSE
          temporalSpatialTradeOffCapability = TRUE
          errorCompensation = FALSE
          maxBitRate = 480
        }
        multiplexParameters = h223LogicalChannelParameters
        {
          adaptationLayerType = al3
          {
              controlFieldOctets = 2
              sendBufferSize = 8192 -- minimum is 1024
          }
          segmentableFlag  TRUE
        }
    }
    reverseLogicalChannelParameters =
    {
        dataType = videoData h263VideoCapability
        {
          qcifMPI = 2
          unrestrictedVector = FALSE
          arithmeticCoding = FALSE
          advancedPrediction = FALSE
          pbFrames = FALSE
          temporalSpatialTradeOffCapability = TRUE
          errorCompensation = FALSE
          maxBitRate = 480
        }
        multiplexParameters = h223LogicalChannelParameters
        {
          adaptationLayerType = al3
          {
              controlFieldOctets = 2
              sendBufferSize = 8192 -- minimum is 1024
          }
           segmentableFlag TRUE
        }
    }
}
```

Figure 7.17   Open logical channel request for a bidirectional logical channel.

```
-- MESSAGE SENT BY PEER INCOMING B-LCSE: B-OLC SUCCESSFUL

Response OpenLogicalChannelAck
{
    forwardLogicalChannelNumber = 2
    reverseLogicalChannelParameters
    {
        reverseLogicalChannelNumber = 2
    }
}
```

Figure 7.18 Bidirectional open logical channel acknowledgment.

The first thing to notice is that the OpenLogicalChannel request of Figure 7.17 includes reverseLogicalChannelParameters, unlike the unidirectional request of Figure 7.12. The request of Figure 7.16 is for forward and reverse H.263 logical channels at QCIF resolution with a frame rate of 15 frames per second and a maximum bit rate of 48 kbit/s, consistent with the declared capabilities of the terminal. The adaptation layer specified is AL3, with a control field of 2 octets and a send buffer size of 8192 octets. The send buffer size allows at least two AL-SDUs to be buffered in case they need to be retransmitted, because maximumAl3SDUSize is set to 4096 in multiplexCapability in the terminalCapabilitySet (see Figure 7.1).

If the receiving terminal accepts the bidirectional open logical channel request it will send an OpenLogicalChannelAck response, as shown in Figure 7.18. The message of Figure 7.18 includes references to both forward and reverse logical channels. The remote terminal uses the Ack to inform the requesting terminal of the logical channel number it has assigned to the reverse channel. In this example it is 2, the same as the LCN for the forward channel.

If the requesting terminal receives an OpenLogicalChannelAck, it sends the OpenLogicalChannelConfirm indication message shown in Figure 7.19 and the process of opening a bidirectional logical channel is complete.

```
-- MESSAGE SENT BY OUTGOING B-LCSE: B-OLC SUCCESSFUL

Indication OpenLogicalChannelConfirm
{
    forwardLogicalChannelNumber = 2
}
```

Figure 7.19 Bidirectional open logical channel confirmation indication.

```
-- MESSAGE SENT BY PEER INCOMING B-LCSE: OLC UNSUCCESSFUL

Response OpenLogicalChannelReject
{
    forwardLogicalChannelNumber = 2
    cause = unsuitableReverseParameters <<null>>
}
```

Figure 7.20 Bidirectional open logical channel rejection.

If the peer terminal does not accept the `OpenLogicalChannel` request for a bidirectional logical channel, it will send an `OpenLogicalChannelReject` response, as shown in Figure 7.20.

In the example of Figure 7.20 the `cause` for the rejection is given as `unsuitableReverseParameters`. Reasons for this could include that the logical channel number is not acceptable because the Multiplex Table Entries of the remote terminal (which have not yet been sent in this example) are incompatible with the `dataType`. This shows that the peer terminal does not like the parameters selected by the requesting terminal for the reverse channel. If this happens the rejecting terminal should immediately send a request to open a bidirectional logical channel with reverse parameters identical to the forward parameters of the rejected request, but with forward parameters that are acceptable to the terminal that initially requested the bidirectional logical channel.

Another reason for failure when attempting to open a bidirectional logical channel is because the other terminal has made a simultaneous (or overlapping) request to open a logical channel for video. This request could be to open either a unidirectional or a bidirectional logical channel. In this case the `cause` will be `masterSlaveConflict` and the terminal that is the master will reject the request from the slave terminal.

If the requesting terminal does not get a response to the `OpenLogicalChannel` request before timer T103 expires, it will send the `CloseLogicalChannel` request of Figure 7.21. Since this is a

```
-- MESSAGE SENT BY OUTGOING B-LCSE: REQUEST TO CLOSE B-LC

Request CloseLogicalChannel
{
    forwardLogicalChannelNumber = 2
    source = lcse <<null>>
}
```

Figure 7.21 Close bidirectional logical channel request.

```
-- MESSAGE SENT BY INCOMING B-LCSE: RESPONSE TO CLOSE B-LC REQUEST

Response CloseLogicalChannelAck
{
    forwardLogicalChannelNumber = 2
}
```

Figure 7.22 Acknowledgment of close bidirectional logical channel request.

request, the incoming B-LCSE that receives it must send a CloseLogical-ChannelAck response.

A similar request will also be sent by a terminal if it does not get the OpenLogicalChannelConfirm message before timer T103 expires. Since this is a request, the terminal that receives it must send a CloseLogicalChannelAck response, as shown in Figure 7.22.

7.1.5 Multiplex table entry exchange

Now that logical channels for audio and video have been opened, all that remains is for the two terminals to exchange multiplex table entries to complete the session establishment. Until this stage is complete the multiplexers at each end will be unable to interpret the MC field in MUX-PDUs carrying media and so will be unable to demultiplex them to extract AL-SDUs.

To start multiplex table entry exchange the terminal sends a MultiplexEntrySend to the remote terminal. An example of this message is shown in Figure 7.23.

The sequenceNumber is incremented with each Terminal CapabilitySet message sent during the session, and is also used in responses, enabling the terminal to match responses to requests. The multiplexEntryDescriptor shows the number of multiplex table entries that are being sent and provides a numbered list of descriptions, each describing a MultiplexElement. In the example of Figure 7.23, four of the possible 15 MIEs are used. The first of these, {LCN1, RC UCF}, allows a MUX-PDU containing audio to be sent and the second, {LCN2, RC UCF}, allows video. The third {LCN1, RC32}, {LCN2, RC UCF} allows MUX-PDUs containing audio and video, where the audio codec is using GSM-AMR at 12.2 kbit/s. The final MTE is {LCN0, RC4}, {LCN2, RC UCF}, allowing an NSRP Ack to be carried along with video information.

```
-- MESSAGE SENT BY OUTGOING MTSE: COMMENCE MUX TAB ENTRY EXCH

Request MultiplexEntrySend
{
    sequenceNumber = 1
    multiplexEntryDescriptors = 4 entries
    {
        [0] =
        {
          multiplexTableEntryNumber = 1
          elementList = 1 entries
          {
            [0] =
            {
              logicalChannelNumber = 1   -- audio
              repeatCount = untilClosingFlag <<null>>
            }
          }
        }
        [1] =
        {
          multiplexTableEntryNumber = 2
          elementList = 1 entries
          {
            [0] =
            {
              logicalChannelNumber = 2   -- video
              repeatCount = untilClosingFlag <<null>>
            }
          }
        }
        [2] =
        {
          multiplexTableEntryNumber = 3
          elementList = 2 entries
          {
            [0] =
            {
              logicalChannelNumber = 1   -- audio
              repeatCount = finite 32      -- one GSMAMR frame at 12.2 kbit/s
            }
            [1] =
            {
              logicalChannelNumber = 2   -- video
              repeatCount = untilClosingFlag <<null>>
            }
          }
        }
        [3] =
        {
          MultiplexTableEntryNumber = 4
          elementList = 2 entries
          {
            [0] =
            {
              logicalChannelNumber = 0   -- allows NSRPAcks
              repeatCount finite 4
            }
            [1] =
            {
              logicalChannelNumber = 2 — video
              repeatCount = untilClosingFlag <<null>>
            }
          }
        }
    }
}
```

Figure 7.23   Multiplex table entry exchange request.

```
-- MESSAGE SENT BY PEER INCOMING MTSE: MUX TAB ENTRY EXCH SUCCESSFUL

Response MultiplexEntrySendAck
{
    sequenceNumber = 1
    multiplexTableEntryNumber = 4 entries
    {
        [1]
        [2]
        [3]
        [4]
    }
}
```

Figure 7.24   Multiplex table entry exchange acknowledgement.

If the remote terminal accepts these capabilities it will respond with the message shown in Figure 7.24, within which the acceptability of each of the MTEs is individually acknowledged.

If the remote terminal does not accept one or more of the MTEs, it will respond with a reject message—Figure 7.25 is an example of a MultiplexEntrySendReject message, showing that the remote terminal did not accept MTE4 with unspecifiedCause. The other possible cause is descriptorTooComplex. This indicates an implementation error; the terminal sending the MTEs should respect the declared capabilities of the remote.

If the remote terminal does not respond before the timer expires, the sending terminal will send an indication that it is abandoning the procedure. This MultiplexEntrySendRelease message is shown in Figure 7.26.

```
-- MESSAGE SENT BY PEER INCOMING MTSE:MUX TAB ENTRY EXCH UNSUCCESSFUL

Response MultiplexEntrySendReject
{
    sequenceNumber = 1
    rejectionDescriptions = 1 entries
    {
        [0]=
        {
         multiplexTableEntryNumber = 4
         cause = unspecifiedCause <<null>>
        }
    }
}
```

Figure 7.25   Multiplex table entry exchange rejection.

```
-- MESSAGE SENT BY OUTGOING TCSE: NO RESPONSE BEFORE TIMER EXPIRY

Indication MultiplexEntrySendRelease
{
    multiplexTableEntryNumber = 4 entries
    {
        [1]
        [2]
        [3]
        [4]
    }
}
```

Figure 7.26 Multiplex table entry exchange release.

Given that the two terminals in our example are identical, they should be able to establish a call straightforwardly without a problem—unless there are excessive transmission errors.

After all the steps we have described so far have completed successfully the two terminals will share a fully established session, with video and audio flowing in both directions.

7.2 Commands and Indications during the Session

During the session the number of H.245 messages sent and received should be relatively low, unless there is some systematic problem, or high error rates. The messages that get sent are most likely to be commands and indications.

A common message that is seen at this stage is the video update command shown in Figure 7.27.

```
-- OUTGOING COMMAND: VIDEO FAST UPDATE

Command MiscellaneousCommand
{
    LogicalChannelNumber = 2
    {
        videoFastUpdatePicture <<null>>
    }
}
```

Figure 7.27 Fast-update command.

```
-- OUTGOING COMMAND: TEMPORAL SPATIAL TRADE-OFF

Command MiscellaneousCommand
{
    LogicalChannelNumber = 2
    {
        videoTemporalSpatialTradeOff = 15
    }
}
-- INCOMING INDICATION: TEMPORAL SPATIAL TRADE-OFF
Indication MiscellaneousIndication
{
    LogicalChannelNumber = 2
    {
        videoTemporalSpatialTradeOff = 15
    }
}
```

Figure 7.28   Spatial temporal trade-off command and indication.

This is a command, so no response is required or expected. It asks the encoder to send an I-frame, which requires significantly more bits per frame than the average, so too many of these commands can actually cause problems and result in poor video quality.

It is recommended that the command to adjust temporal spatial trade-off is responded to by an indication of the latest setting for the remote encoder. Figure 7.28 shows this command and the indication that follows.

7.3   Terminating the Session

The session is terminated when one or other of the terminals issues the end session command shown in Figure 7.29.

This leads to an abrupt end to the session, there is no requirement to go through any controlled process of cleanly closing logical channels.

```
-- OUTGOING COMMAND: END SESSION

Command EndSessionCommand
{
    Disconnect <<null>>
}
```

Figure 7.29   End session command.

7.4 Chapter Summary

The video telephony call that has been presented in this chapter has been intended to illustrate and reinforce the concepts presented in earlier chapters, consolidating the information and giving specific examples of the syntax and semantics of H.245 messages.

Although not every procedure has been walked through, by using the information supplied in this chapter along with Chapter 6 and Appendes A and B you should be in a position to deduce what the messages for procedures that have not been explicitly illustrated will look like. You should by now have a good basic understanding of how a 3G-324M terminal operates.

Chapter 8

Implementation Issues

Many aspects of a 3G video telephony terminal, piece of network equipment, or end-to-end video telephony service are specified by the various recommendations that make up 3G-324M. It should be clear from previous chapters that 3G-324M is fairly complex. This complexity means that ensuring compliance with the standard is far from trivial.

Diagnostic and testing tools are essential for the development of systems and the deployment of services. Diagnostic tools can assist in determining compliance by capturing the data streams that are exchanged when a call is made and analyzing them to extract the MUX-PDUs, AL-SDUs, and H.245 messages. Testing tools allow comprehensive testing to be performed, exploring the compliance of the system under test, exploring the boundaries of its compliance, and checking its response to noncompliant terminals.

Whilst much of a 3G-324M terminal or piece of network equipment is specified by standards, there are many areas that offer flexibility and the opportunity for implementers to add value and differentiate their products from those of their competitors.

This chapter covers implementation issues. It highlights areas where noncompliances can arise, discusses how systems can be diagnosed and tested, and suggests where implementers can use their design expertise to build systems that perform well.

8.1 Conforming to Standards

The importance of 3G mobile video telephony terminals and network equipment conforming to standards cannot be overstated. If the early terminals offered by mobile operators deploying video telephony services on 3G networks are not fully compliant to 3G-324M, a potentially significant population of users of these terminals could build up. Should this happen it would create a serious problem to which there are only two solutions. The handsets could be recalled and replaced with handsets that are compliant, or it could be made a requirement that later compliant handsets can also interwork with the legacy noncompliant handsets. Both of these solutions are undesirable and expensive.

Internet multimedia terminals based on H.323 or SIP are generally implemented in software on PCs; if a noncompliancy is discovered in a particular release of software, an update can easily be downloaded. Although advances are being made in this area, the firmware in a mobile phone is not easily upgraded in the way that a PC client can be. Because of this, rigorous interoperability testing needs to be performed prior to the release of a 3G-324M handset. This must include comprehensive testing against products from other manufacturers if at all possible. No manufacturer will release a product that fails to interwork with other examples of the same product, or other products in its portfolio; this does not mean that it will necessarily work reliably with products from other manufacturers.

The International Multimedia Telecommunications Consortium (IMTC) is an industry organization whose aim is to encourage the widespread adoption of multimedia services. It does not produce standards, though it liaises with standards-making groups. One of its aims is to encourage interoperability testing. It has an activity group dedicated to 3G-324M, which holds regular interoperability testing events providing an opportunity for member organizations to test against each others products.

The work of the IMTC forum is extremely valuable, but its schedule of interoperability testing is unlikely to align well with particular product development timescales. The scope of the testing is dependent on partners and therefore not within the control of an individual company. It is therefore important that companies working in this area have their own capability to perform extensive interoperability testing. This applies equally to manufacturing companies developing terminals or network equipment and to operator companies deploying services that make use of components from several different manufacturers.

8.2 Diagnosing Problems

If we are able to look at the bit streams, the extracted MUX-PDUs, the AL-SDUs, and the H.245 messages flowing between terminals, we should be able to diagnose most problems that can arise in a 3G-324M call.

The most systematic approach to analyzing problems in 3G-324M equipment is to go through each stage of the call and session establishment process and look at what can go wrong.

8.2.1   Bearer establishment

The establishment of the bearer was discussed briefly in Chapter 4. There needs to be some 3G-324M awareness at the bearer level. The signaling must indicate that a transparent (UDI) channel is required, with User Information at layer 1 set to H223&H245 and LLC bits 1-5 set to 00110 to indicate a multimedia call. If not the call may well be rejected.

Checking and diagnosing problems at this level can be done with readily available network protocol analyzers.

8.2.2   Synchronizing multiplexers

Once the bearer is established, the first area where problems can arise is when the 3G-324M terminals attempt to synchronize by sending stuffing sequences and examining the stuffing sequences they receive.

The wrong stuffing sequences may be sent, or the sequence may not be sent for long enough. Alternatively terminals may wait to see what stuffing sequence they receive from the other before starting to send a stuffing sequence, resulting in no activity. In the case of calls over a live network as opposed to a laboratory network model, error rates on the connection may be too high to allow synchronization.

The cause of the problem should be apparent by examining the bit stream to check that flags are being produced by both terminals. If they are, their validity and duration can be checked.

8.2.3   MUX-PDUs

Once the multiplexers have synchronized, each terminal should be able to extract MUX-PDUs from the incoming bit stream and H.245 messages should start to flow in LCN0. If the call fails at this stage without the terminal successfully detecting MUX-PDUs, this could point to a problem with opening and closing flags. This is unlikely because they are used in the stuffing sequences for multiplexer synchronization.

Another possibility is that the received MUX-PDUs may be too large for the terminal. The bit stream can be examined to see if it contains anything other than stuffing sequences, to check that the terminals are attempting to send information.

8.2.4   AL-SDUs

If the terminal detects MUX-PDUs in the bit stream but no H.245 messages are decoded, the problem may be at the adaptation layer or the NSRP or CCSRL layers. This can be determined by looking at the AL-SDUs, in particular the NSRP Acks to see if they are correctly formed; if so, the problem may be at the CCSRL layer, or with the PER encoded H.245 messages themselves.

8.2.5   H.245 procedures

If H.245 messages are successfully extracted, diagnosis of any problems is made easier by looking at the H.245 messages, provided that they are available in human-readable rather than PER encoded form.

General problems with H.245 procedures may include timer related problems, problems with the sequencing of messages, and problems with unexpected messages.

If the timers in either terminal expire too quickly then procedures may frequently fail. A clue that this is a problem will be the frequent occurrence of release indications from the outgoing SE, and repeated attempts to initiate some procedures. The problem may be confined to one direction of the call if different equipment is used at each end.

If the timers are set to values that are too large, session establishment times may become very long in conditions where the level of transmission errors means that procedures such as capability exchange and master slave determination need to retry.

If a terminal is too rigid in its implementation of the sequencing of procedures and does not accept the order in which the other tries to establish the session, this may cause problems. Terminals may also behave unpredictably when presented with a message that they do not themselves support, rather than responding with a FunctionNotSupported indication where appropriate, or simply ignoring the message if it is an indication.

8.2.6   Capability exchange

The first H.245 message sent should be a capability set request. If a terminal has a problem with the capabilities that are sent, it should reject them. The cause value in the reject message should indicate the reason for the rejection. Causes may include that the sending terminal has sent more capability descriptors or more table entries than the terminal can handle. This type of issue would point to a failure to comply with the standard.

Problems with capabilities may not be apparent immediately. The procedure may complete successfully only for capability-related issues to occur later. The TerminalCapabilitySet message specifies the declared capabilities of each terminal under test. Once this is captured as part of any test procedure, it can be used to plan further tests to explore the conformance of the terminal to the capabilities that it has declared, and to check that it respects the capabilities of other terminals.

8.2.7   Master/slave determination

In the master/slave determination procedure, terminals may not respond correctly to terminal type values other than 128, they may incorrectly assign status leading to later problems with such procedures as opening logical channels. These problems will probably not be experienced in a call between terminals from the same manufacturer, but will become evident in calls to other terminals.

If the number used to determine status is not truly random, the probability that the procedure will initially fail and be forced to retry is higher.

8.2.8   Exchanging multiplex table entries

The terminal may itself use less than the maximum of 15 MTEs that are specified in H.223. This is perfectly acceptable, provided that it will accept the maximum number of MTEs if they are sent to it by another terminal.

A terminal may have declared support for enhanced MTEs in its capabilities but reject them when they are sent as part of the `MultiplexEntrySend` message, or fail to use them correctly when demultiplexing.

8.2.9   Opening logical channels

The behavior of terminals when opening logical channels is another area where problems may be observed; logical channels may fail to open properly.

If the peer terminal does not accept the parameters selected for the reverse channel of a bidirectional logical channel by the requesting terminal, it should reject the request and immediately send a request to open a bidirectional logical channel with reverse parameters identical to the forward parameters of the rejected request, but with forward parameters that are acceptable to the terminal that initially requested the bidirectional logical channel.

If terminals make overlapping requests to open logical channels, this can result in a conflict. This applies to both unidirectional and bidirectional logical channels. Such conflicts should be resolved by the master rejecting the request from the slave terminal.

Where overlapping requests are made to open a bidirectional logical channel and the intention is that there should be just one bidirectional video channel the master terminal must reject the request from the slave.

Cases of conflicting unidirectional channels should not arise if the terminals follow the recommendations in H.245 on declaring their preferences in the terminal capabilities and selecting the media type for such channels. This needs to be tested because it is recommended rather than mandatory behavior. If a conflict does occur the terminals should resolve it through the master terminal rejecting the request from the slave.

It should also be confirmed that in such cases the slave terminal will then request a unidirectional logical channel for a nonconflicting media type.

8.2.10   Problems with media

The session may reach the point where logical channels for audio and video are open, only for there to be problems with the media being transmitted and received.

If no video is seen or audio is heard, the reception of AL-SDUs for the media type should be checked. If they are received, there may be a problem associated with picture start codes for H.263 or visual object headers for MPEG4—for example they may not be placed in the AL-SDUs correctly to align with the start of an AL-SDU. For audio, the number of frames in each AL-SDU could be checked to check that it corresponds to what has been declared in the open logical channel request. If the AL-SDUs look valid there may be a compatibility problem between the remote encoders and the decoders in the terminal experiencing the problem.

It is always worth checking the volume and contrast settings!

If video (or audio) is present but noisy or corrupted, the reason may be high transmission errors: frequent video fast update commands indicate that the video decoder is having difficulty. If the terminal issues too many of these, it can contribute to rather than ameliorate the problem.

If the optional maintenance loop procedure is implemented, it can be used to help examine such problems for bidirectional channels over AL3.

8.2.11   Other procedures, commands, and indications

Other procedures of the 3G-324M terminal, such as switching mobile levels, mode requests, closing logical channels, and responding correctly to temporal spatial trade-off messages all need to be exercised to check that they function correctly, and the terminal must be checked to see that it is well-behaved when it receives an H.245 command or indication that it does not support.

8.3   Diagnostic and Testing Tools

Diagnosis is concerned with determining what is going wrong; testing is concerned with exploring the limits of the device or system under test.

By following the approach outlined in the previous section a comprehensive test suite can be built up to check that a terminal functions correctly and in accordance with relevant recommendations. Running this suite requires test equipment that can be flexibly configured to probe the limits of a device.

A handset manufacturer may extract some diagnostics when a terminal is in development, but once a handset is shipped as a product it is unlikely to provide any diagnostic capability. It is also unlikely that it will allow a user to selectively exercise its functionality, beyond entering the destination phone number and pressing the call button. Because of this it can be difficult to diagnose problems experienced when 3G-324M handsets from different manufacturers are used to call each other, or to call a piece of 3G-324M network equipment, without third party tools.

Very few diagnostic tools are currently available, and even fewer test tools. The only commercial tool that combines on-line real time diagnosis and test capabilities is the Dilithium Networks Analyzer (DNA). This is a very feature-rich tool that has become the industry standard tool for testing 3G-324M devices and systems. In case I am suspected of any self-interest in saying this, I suggest that you do a search on the web using such keywords as 3G-324M, test, tool, and the like. The few tools that may be available from other sources are restricted to off-line analysis of bit streams captured during a call, and provide diagnostic assistance rather than the ability to perform proactive testing and exercise the equipment under test in unexpected ways.

DNA operates in two modes. It can be used as one endpoint of a call or it can be used to noninvasively monitor a call. For diagnostic purposes it allows simultaneous logging of bit streams, MUX-PDUs, AL-SDUs, received media data, and H.245 messages. This logging can be performed at the same time that the operator views the video and listens to the audio.

For testing purposes DNA can be configured with a series of profiles that allow a wide range of terminal capability sets to be defined. MTEs can be specified by the user and command and indication messages can be manually triggered.

The monitoring mode is particularly useful for mobile operators to investigate and diagnose problems in their networks, and to look at problems that arise in the interaction between handsets or other equipment from different manufacturers.

8.4 Design Considerations

A 3G-324M handset must comply with relevant standards. This still leaves many areas where good design and implementation can differentiate an excellent product from a merely adequate one.

Referring to Figure 2.15, the obvious place for product innovation is in the areas that are out of scope of the standards. These include the user interface; the quality of the media capture and play out devices; the physical characteristics, such as size, weight, and robustness; and battery life.

A higher quality lower noise camera or microphone will provide a cleaner source signal. This results in more efficient encoding of the signal and therefore better quality though the beneficiary of this will be the person at the far end of the call. Higher quality displays and speakers are also better from the perspective of the user.

Innovative user interface design that makes using the handset intuitive is also important. Release 99 3G-324M phones have two separate buttons: one for making video calls and the other for making speech calls—these should be clearly distinguishable.

The SE user function is largely outside the scope of multimedia standards, apart from some loose, high level constraints on the sequencing of the operations it performs. In those areas of the handset that are within the scope of the standards there are also aspects where good design can lead to a better quality product.

8.4.1   SE user design

The SE user is itself a state machine that reflects the state of the session. It holds the information about such things as the capabilities and the MTEs of the local terminal and provides this information to the SEs. Each SE informs the SE user of the content of almost all incoming messages. The SE user forwards information to components such as the multiplexer and the codecs as needed. When the SE user receives information on the capabilities of the remote terminal it uses it to decide on the logical channels to be opened; it instructs the SEs to send request messages and makes decisions on the response messages to be sent.

The intelligence of the terminal is in the SE user: the SEs are its dumb agents. The degree of robustness and flexibility that a terminal exhibits is very much dependent on the SE user. A well designed SE user will be tolerant of behavior on the part of the remote terminal that is not expected. For instance, if hypothetically a remote terminal initiated master/slave determination before terminal capabilities exchange, this would be noncompliant, but does not in principle mean that session establishment must fail. A well designed SE user component should show as much tolerance of variations in the sequencing of procedures on the part of the remote terminal as is possible without compromising the possibility of successfully establishing or maintaining the session.

Other areas that are left to the implementer include the design of MTEs. There is scope for creativity in this area: the SE (and multiplexer) could have several alternative sets of MTEs. The SE user could decide which set to send only after seeing the capability set of the remote terminal, allowing it to choose between sets of MTEs optimized for different combinations of the video and audio codecs that it supports.

Deciding on the circumstances in which to initiate changes to the multiplexer configuration is also left to the implementer. A careful choice of a set of error metrics to use to initiate a change of *mobile levels* or optional modes (such as ML1 with double flag or ML2 with optional header if these options are supported) may result in better performance in different transmission error conditions. The benefits of this may, however, be marginal given that the multiplexers start at their highest common mobile level.

For some procedures, such as capability exchange, it is recommended that if the process fails it should be re-tried. There is no reason why when the process is retried the same terminal capabilities should be used— retrying with a more conservative set of capabilities may enable a successful capability exchange.

For procedures that do not explicitly specify that retries should be attempted, retrying is not disallowed and there may be benefits in doing so. If a logical channel fails to open as a result of a local timer expiry attempting a retry may be a better strategy than giving up immediately.

In these and other ways the designer of a 3G-324M handset, or of other 3G-324M compliant equipment, can add value to their product and differentiate it from competing products.

8.4.2   Design of 3G-324M components

There is scope for good design (as opposed to implementation efficiency) of those components that fall within the scope of multimedia standardization. Good design in this area can mainly be categorized as resulting in performance improvements.

The design of MTEs that allow efficient transport of the media and control data has already been mentioned, as has optimization of timer values.

A well designed 3G-324M terminal should be optimized to maximize its use of the available bandwidth without risking running into problems with loss of data because the video and audio codecs are producing more encoded information than can be sent, or on the other hand sending insufficient information causing wasteful stuffing sequences to be transmitted. Audio AL-SDUs are produced at a constant rate of 33 per second for G.723.1 or 50 per second for GSM-AMR, with the number of octets per AL-SDU dependent on the mode. The rate at which the video codec generates information is the variable that can be controlled to achieve optimum usage of available bandwidth.

A 64 kbit/s channel is equivalent to 8000 octets per second. Once the channel is established there will be very few messages in LCN0, so if we assume that there is one audio AL-SDU per MUX-PDU we can calculate that a suitable MUX-PDU size is about 240 octets if G.723.1 is being transmitted, and 160 octets if GSM-AMR is being transmitted. It is recommended by 3GPP that MUX-PDUs are limited to less than 200 octets to avoid transmission errors, so it may be advisable to aim for160 octet MUX-PDUs in both cases and expect that single G.723.1 audio frames will be put into in two out of three MUX-PDUs. The overhead associated with the adaptation layer types and mobile level in use is known, so the remaining octets available for video can be estimated. The video codec should be controlled to operate just within this limit, and MTEs must be provided that support the expected mix of MUX-PDUs.

The algorithms to be used for dynamically creating MUX-PDUs from incoming AL-SDUs are not specified in the standards and provide plenty of scope for design innovation.

Setting correct timer values can have a significant impact on performance in the early stages of the call, during session establishment. The ideal value is one that is just a little longer than the round trip delay, which is partly network dependent. Executing the round trip delay procedure early in the call and using the measurement to adaptively adjust the timer values is a possibility. There is unlikely to be any significant benefit in using different timer values for different processes, except for the round trip delay process itself, which should have a timer value set to a reasonable upper limit of perhaps two to three times the expected round trip time.

8.5   Implementation Considerations

The implementation of a 3G-324M handset is dependent on the hardware selected. The target processor may be a *general purpose processor* (GPP) with codecs running on an *application-specific integrated circuit* (ASIC) or there may be a general purpose processor such as an ARM RISC processor and a *digital signal processor* (DSP) chip available. We will consider the latter case.

The GPP and the DSP may be separate processors, or they may be dual cores on a single piece of silicon. The Texas Instruments (TI) OMAP processor is an example of a chip targeted at low power mobile applications. It includes an ARM core and a TI C55xx DSP core and supports a range of operating systems including Symbian. When considering how a 3G-324M system can be mapped onto such a platform, it is clear that the codecs with their repetitive signal processing computations are suitable for the highly pipelined architecture of the DSP. H.245 and the SE user involve considerable conditional branching and are therefore much better suited to implementation on the GPP. The multiplexer is probably best suited to implementation on a DSP architecture, but the decision eventually depends on considerations such as which processor reads from and writes to the network interface and which processor has capacity available.

If the flows of information in the system are considered, by far the highest data transfer rates are required between the video codec and the camera and display devices, where uncompressed video must be transferred, so direct communication between these devices and the DSP is strongly indicated. Uncompressed audio data, the multiplexer input and output streams, and the combination of encoded video and audio are each about 64 kbit/s, which is not a demanding amount of data to transfer between processors if the multiplexer is placed on the GPP. If the DSP handles the network interface and the multiplexer is placed on the DSP, the interprocessor communication reduces to the transfer of H.245 messages. If alternatively the GPP handles the network interface, interprocessor transfer of the 64 kbit/s incoming and outgoing multiplexed streams is also required.

Irrespective of how the 3G-324M stack components are partitioned, resources such as computational power and available memory will be limited, to keep overall system power consumption low and maximize battery life. Handset manufacturers are keen to put as many features and functions into their products as possible, so by no means all the resources of the handset are available to the 3G-324M stack. The computation required for the video and audio codecs is much larger than the computational demands made by the multiplexer but the limited resources of the DSP will still demand a highly efficient implementation of the stack components.

The design challenge is to make the implementation as efficient as possible while maintaining good performance, measured by efficiency of bandwidth usage and session establishment setup times.

A number of implementation decisions can influence computational and memory requirements. There are opportunities for making trade-offs between the two in the implementation of H.245 and the multiplexer, depending on which is in shortest supply. As an example, CRCs can be generated and checked using algorithms that make extensive use of lookup tables to reduce computation, or alternatively they can be fully computed to reduce lookup table sizes and therefore memory requirements.

Manufacturers usually choose to significantly lower their development risks by licensing proven third party implementations from suppliers such as Dilithium Networks. These may be available on the target platform and provide the manufacturer with the reassurance that they have been extensively tested for interoperability. This approach shortens development lifecycles and avoids the otherwise significant risk of noncompliance with the many and complex standards that comprise 3G-324M if development of a stack in-house is attempted.

8.6 Chapter Summary

In this chapter the importance of conformance to standards has been emphasized. Products from different manufacturers must be fully interoperable if the deployment of 3G video telephony is to succeed. We have looked at how a wide range of problems can be diagnosed, and the tools needed to diagnose and test 3G-324M systems.

The recommendations of 3G-324M are not so prescriptive that they stifle design innovation. Some of the areas that offer scope for creative design have been discussed, and some of the aspects that need to be considered in implementing the designs have been reviewed. The emphasis has been on handsets, but much of what has been discussed is equally applicable to network equipment and other systems that support 3G video telephony.

9

Video Telephony over Mobile Packet Networks

The evolution of 3G networks after the initial deployment of Release 99 was discussed in Chapter 1. Release 99 provides circuit-switched access for video and speech telephony and packet-based access for Internet and data applications. Later releases migrate to a completely packet-based infrastructure, with UE to network signaling based on the use of the *session description protocol* (SDP).

In Release 99 based networks mobile speech terminals and video telephones are distinct and do not have a way of directly interworking. Any handset that incorporates video telephony will also include a speech client, and the user of the device will have to indicate the type of call they are requesting by selecting the correct call button. In contrast, SIP terminals with different media capabilities will interwork seamlessly after the capabilities are established at the start of the call, provided they have some overlapping capabilities. The *session description protocol* (SDP) within the payload of SIP messages is used by the terminals to inform one another of their capabilities and agree the media streams and formats to be transmitted. The media is then streamed using RTP.

In this chapter we will look at these protocols and at how they are used in 3G packet-based video telephony terminals.

9.1 The Session Initiation Protocol

Multimedia terminals over the 3G packet network use IETF RFC 3261 "SIP: Session Initiation Protocol" for session establishment and control.

A SIP terminal is known as a *user agent* (UA). The UA is made up of a *UA client* (UAC) that issues SIP requests, and a *UA server* (UAS) that responds to requests.

A direct UA to UA call is possible but to offer a scalable SIP-based service servers are needed in the network. There are a number of different types of network-based servers. These include registrars, location servers, and proxy servers.

The function of a registrar is to associate an SIP address, known as an *SIP uniform resource locator* (SIP URL), of the form SIP:user@host.domain, with an IP address. When a SIP UA is first powered up or started, it will communicate with the registrar to authenticate itself and establish the mapping of its SIP URL to an IP address. The packet networks for 3G Release 5 and later releases are based on the use of IPv6, with a 128-bit address instead of the 32-bit address used in IPv4. There is already an acute worldwide shortage of available IPv4 addresses. IPv6 provides a much greater address space and so is able to satisfy the expected demand for IP addresses that will be generated as IP-based 3G mobile networks are deployed. IPv6 addresses are of the form AAAA:BBBB:CCCC:DDDD:EEEE:FFFF:GGGG:HHHH, where each letter represents a hexadecimal value.

SIP location servers store information on where users can be found. A location server may be updated by a registrar so that it holds the latest information on registered users.

Proxy servers are used to discover the location of the called party, set up calls, and to route messages between the endpoints. When a call request is made, the proxy server that receives the request may communicate with a location server to determine where the request should be forwarded to. In practice there may be more than one proxy server in the signaling path to establish a call. The presence of the proxy server in the signaling path also enables call records to be captured for billing.

9.1.1 SIP messages

SIP messages are human-readable and are based on http. Many, but not all, SIP messages can contain SDP messages in their payload or message body.

SIP messages are either requests or responses. The core SIP requests, known as methods, are REGISTER, INVITE, ACK, OPTIONS, BYE, and CANCEL:

REGISTER is used by the UA to register, to provide the details of its current IP address to a registrar.

INVITE, as its name implies, is a request used to invite someone to participate in a session, and is effectively a call request. An INVITE request may result in a number of responses, indicating the progress of the network equipment (proxy servers) and peer UA in determining the outcome of the INVITE request.

ACK is used to acknowledge the reception of a final response to an INVITE request, whether successful or not. No response is generated to the ACK request. This three way INVITE/final response/ACK handshake is intended to provide more reliable session establishment.

OPTIONS enables a UA to obtain information about the capabilities of a remote UA or proxy server.

BYE is used to leave a session. For two party sessions this effectively terminates the session.

CANCEL cancels any current request transactions that have not completed.

The responses that may be generated as a result of receiving a request are numbered between 100 and 699 with an associated text-based reason. They fall into classes based on the first digit of the range that the number used falls into. Table 9.1 shows some of the classes of responses, with examples of specific responses. The similarity to http response codes is clear.

Informational messages are used to keep the UA informed of the progress of a request. Success messages indicate that the outcome of the request was successful. Error and failure responses are sent when requests cannot be fulfilled for any reason.

An INVITE message sent by a UAC can cause a number of provisional responses to be generated by the peer UAS before the final response is sent. Timers and a retry mechanism are used to ensure reliability of the transaction initiated by sending the requests. There are some differences in the way this is handled for INVITE, but in general the request is retransmitted if no response is received, at intervals that double for each retransmission up to six times. The initial interval should be set approximately equal to the round-trip time.

TABLE 9.1   Some Examples of SIP Response Messages

Response class	Number range	Examples
Informational	100–199	100 Trying 180 Ringing 183 Session progress
Success	200–299	200 OK 202 Accepted
Client error	400–499	404 Not found 408 Request time-out 486 Busy here 487 Transaction cancelled
Global failure	600–699	604 Does not exist anywhere

SIP extensions can be used to provide further request methods in addition to this core set. Examples are PRACK and UPDATE. PRACK, defined in RFC 3262 "Reliability of Provisional Responses in the Session Initiation Protocol," is a request that is used to provide provisional acknowledgments in some circumstances. UPDATE, defined in RFC 3311 "The Session Initiation Protocol (SIP) UPDATE method," is a means of revising an SDP session description that was originally contained in an INVITE request, before the final response has been received and the session has been established. UPDATE makes use of PRACK for reliability.

9.1.2   An example of a call using SIP

A simple call establishment procedure using SIP is shown in Figure 9.1.When a SIP call is made the UA will generate an INVITE request to make a call to another user. The INVITE request is sent to the proxy server. The proxy server interacts with the location server to determine the IP address of the called party and uses this to modify and forward the INVITE request to the destination.

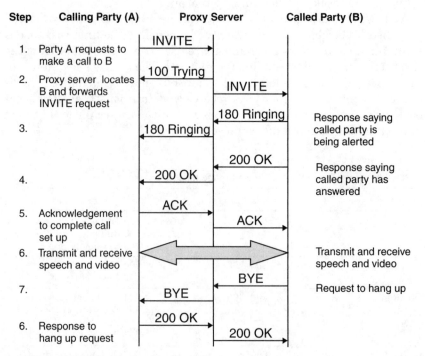

Figure 9.1   Making a simple call with SIP using a single proxy.

```
INVITE sip:partyB@mobiletelco.net SIP/2.0
Via: SIP/2.0/UDP [aaaa.bbbb.cccc.dddd.eeee.ffff.gggg.hhhh]:5060
From: partyA <sip:partyA@mobiletelco.net>
To: partyB <sip:partyB@mobiletelco.net>
Call-ID: 18273645@proxy.mobiletelco.com
Cseq: 1 INVITE
Contact: <sip:[aaaa.bbbb.cccc.dddd.eeee.ffff.gggg.hhhh]:5060>
Content-Type: application/sdp
Content-Length: 216

{message body}
```

Figure 9.2   INVITE request from party A.

Although we will not go into the syntax and semantics of SIP messages in any depth here, we will look at a couple of examples related to the call of Figure 9.1. Figure 9.2 is a listing of the INVITE request sent by party A to initiate the call.

This message has a number of features. The first line is the request line, showing that it is an INVITE request, and containing the SIP address of the called party. The second line shows that the call is being made using SIP version 2 over UDP from the IP address specified (using port 5060). The call-ID is used by the UA to distinguish this call from other calls that may be taking place between the same participants.

The Cseq, or Command Sequence, line identifies unique requests associated with the Call-ID. This allows responses (which include the Cseq number within them) to be uniquely associated with requests, avoiding ambiguity. All requests following an initial INVITE request have Cseq incremented, except for ACK, which uses the same Cseq value as the INVITE response it is acknowledging, and CANCEL, which uses the Cseq value of the INVITE request it is attempting to cancel.

Wherever URLs are used, they can be replaced by a corresponding IP address. In the Contact line the IP address of the calling terminal is specified, so the called terminal is in principle able to send responses directly to the calling terminal rather than via the proxy(s).

The Content-Type specifies what is in the payload or body of the message, in this case it is SDP content. The Content length shows the number of bytes that are in the payload. The use of SDP in the payload is discussed later in this chapter. SIP messages can contain more information than is shown in this example: we have reduced the information in the request to keep the example simple.

When the proxy server receives the INVITE message of Figure 9.2, it can manipulate the header in various ways. For the simplest case where both UAs are in the same domain as the proxy, it looks up the IP address of the destination and uses it to replace the URL of the called party in the first line of the INVITE request. It may insert extra Record-route lines

```
SIP/2.0 200 OK
Via: SIP/2.0/UDP [ssss.tttt.uuuu.vvvv.wwww.xxxx.yyyy.zzzz]:5060
From: partyB <sip:partyA@mobiletelco.net>
To: partyA <sip:partyB@mobiletelco.net>
Call-ID: 18273645@proxy.mobiletelco.net
Cseq: 1 INVITE
Contact: <sip:[ssss.tttt.uuuu.vvvv.wwww.xxxx.yyyy.zzzz]:5060>
Record-route:<proxy.mobiletelco.net:5060>
Content-Type: application/sdp
Content-Length: 197

{message body}
```

Figure 9.3 200 OK response from party B.

in the header, which force the called terminal to send its responses back via the proxy rather than directly. This enables the proxy server to follow the progress of the call and gather data for billing and other purposes.

The proxy server may send informational responses such as 100 Trying or 183 Session Progress to the calling terminal whilst doing this, to ensure that it knows that its request is being acted on.

When the called terminal receives the modified INVITE request it alerts the user, and sends the informational response 180 Ringing to the proxy, which is forwarded to the calling terminal.

When the user answers the call, the success response 200 OK is sent. The 200 OK response is shown in Figure 9.3.

This response reflects much of what was included in the header of the INVITE request. The addresses on the From and To lines have swapped as expected, the Call-ID and Cseq values identify it as a response to the INVITE sent earlier. The Contact line contains the IP address for Party B. The Record-route line is included because it was inserted by the proxy when it forwarded the INVITE request.

The proxy forwards the 200 OK response to party A, which completes session establishment by sending an ACK request. Based on whatever has been agreed using SDP, which we will discuss next, media can now be sent between the terminals. When either party wishes to end the call, they send a BYE message generating a 200 OK response from the other terminal.

SIP headers can contain many other lines and fields; the simple examples we have looked at are only illustrative.

9.2   The Session Description Protocol

SIP provides a means of starting and ending sessions but on its own it says nothing about what will happen in the session once it is established. SIP terminals send descriptions in their payloads using RFC 2327 "SDP:

```
v=0
o=partyA 2890844526 2890844526 IN IP6 partyA.mobiletelco.com
s=-
c=IN IP6 [aaaa.bbbb.cccc.dddd.eeee.ffff.gggg.hhhh]
b=CT:128
t=0 0
```

Figure 9.4 First part of SDP example.

Session Description Protocol" to exchange capabilities and agree on the media types that they will send to one another.

SDP provides a way of describing the characteristics of a terminal. SDP session descriptions comprise lines of text with the syntax <type> = <value>, where <type> is always exactly one character and is case significant. <value> is a structured text string whose format depends on <type>. Some lines are mandatory and some are optional but if present they always appear in a fixed order.

A session description has one or two parts. The first part describes the session as a whole. The second part (if present) provides descriptions of the media that are supported. An example of the first part of a session description is shown in Figure 9.4. This does not show all the lines that can be in this part of the description, but includes mandatory lines and optional lines of interest in 3G mobile video telephony.

Taking the example of Figure 9.4 line by line, the first line is the version number and is always set to 0. The second line specifies the originator of the message and provides a user name, session identifier, a version, network type, address type, and address. The third line is the session name. For a point to point video call this is not needed and so is set to "–". The fourth line provides connection information, specifying a destination or source address for any media in the session description. The fifth line is an optional line that provides a bandwidth limit in kbit/s for the overall session (CT = Conference Total). The connection and bandwidth lines in the session-level information of Figure 9.4 apply to all the media in the session unless they are overridden in specific media descriptions. The last line of this part of the session description gives the start and stop times. This is more relevant to the original purpose of SDP to make announcements of multi-party sessions. In our case it is set to "0 0".

Figure 9.5 gives an example of the second part of the session description.

Referring to Figure 9.5, the lines beginning with m are media descriptions. These specify the media type the description relates to, the UDP port number the media will be sent to, the way the media is transported, which in all cases we are interested in uses RTP with *audio visual profile*

```
m=audio 48068 RTP/AVP 96 4
a=rtpmap:96 AMR/8000
a=rtpmap:4 G723/8000
m=video 49100 RTP/AVP 34 97
b=AS:112
a=rtpmap:34 H263/90000
a=rtpmap:97 MP4V-ES/90000
a=fmtp:97 profile-level-id=1
```

Figure 9.5   Second part of SDP example.

(AVP) as specified by RFC 3551 "RTP Profile for Audio and Video Conferences with Minimal Control." Following this are one or more numbers, which are known as payload types (PTs). When there is more than one it means that more than one format can be supported. The order in which the PTs are listed indicates their order of preference. PTs can be static or dynamic. Static PTs have numbers less than 96. The range 96–127 is used for dynamic PTs. The reasons for this are historical: initially video and audio standards were assigned static PT numbers. Examples are G.723.1 that is assigned to 4 and H.263 that is assigned to 34. With the proliferation of codecs and varying profiles within codecs, it was considered that this approach was not scalable, so newer standards and codecs are registered as *multipurpose internet mail extension* (MIME) types (RFC 3555 "MIME Type Registration of RTP Payload Formats") and have PT numbers dynamically assigned for the duration of a session, with any additional required information specified in the attribute lines that follow it. The "a=rtpmap" attribute provides the binding of the PT number to a particular media format, for dynamically assigned PTs. Although we have specified this attribute for PT 4 in our example, it is not strictly needed for static PTs.

The "a=fmtp:" attribute is used to provide parameters specific to a particular format. In the example it specifies that PT 97, MPEG4, uses Simple Profile Level 0.

Other attributes of interest include the following:

- a=ptime: <packet time> gives the time in milliseconds of the media in a packet and can be used to specify how many encoded audio frames are placed in a single RTP packet.

- a=recvonly indicates that media can only be received.

- a=sendonly indicates that media can only be sent.

- a=sendrecv specifies that the media should be started in send and receive mode and is analogous to the ReceiveAndTransmit capability in H.245. It is usually unnecessary to specify this, because it is the default unless a = recvonly, or a = sendonly are explicitly included.

- a=inactive indicates that no media is sent
- a=framerate:<frame rate> gives the maximum video frame rate in frames/s.
- a=quality:<quality> indicates the quality of video; the value is in the range 0 to 10, where 0 is the lowest quality and 10 is the highest.

In the example of Figure 9.5, the session description offers audio in either GSM-AMR format (preferred) or G.723.1 format, and video in either baseline H.263 (preferred) or MPEG4 SP@Level0. The maximum bandwidth of the video is specified in the line following the media description; b = AS:112 as 112 kbit/s, where AS stands for Application Specific. Setting the bandwidth to zero indicates that no media should be sent. It differs from the a=inactive attribute in the way it is dealt with by RTP and RTCP.

9.2.1 The offer/answer model

Terminals can use session descriptions in SIP messages to perform the equivalent of capability exchange in H.245 and agree on what media is going to be exchanged in the session. This is defined in RFC 3264 "An Offer/Answer Model with the Session Description Protocol."

The offer/answer model is used during the establishment of a session to agree on the type and format of media streams that will be exchanged during the session. The calling party can make an offer to the called party in the form of a session description containing the media streams and formats it wants to use, and other relevant information such as the IP address and port numbers it wants to receive the media streams on.

The offered session description can be sent in the SIP INVITE request. If the called party is unhappy with the session description as a whole it can simply reject the INVITE request with a response such as 606 not acceptable.

If the called party is able to respond to the offer, it sends an answer. This is a session description that may be contained in the 200 OK response or in earlier informational responses such as 183 Session progress.

The offer/answer model can also be used by either party to modify an active session, for example by modifying frame rate, quality, or other attributes, provided that it is not waiting for an answer to an earlier offer and it has not received an offer to which it has not yet responded.

The initial offer contains media descriptions and any associated attributes. The example of Figures 9.4 and 9.5 could be the session description for an offer. If the calling party wants to agree on the media types and formats to be used but doesn't want to send or receive media initially it can add the attribute "a=inactive" after each media description.

If multiple formats are listed in the media description, it means that the party sending the offer is capable of making use of any of those formats during the session, listed in decreasing order of preference. By analogy with H.245, each media description in the offer represents a simultaneous capability and each format listed within the media description represents an alternative capability. The only difference is that multiple formats in a media description imply that a change can be made between any of the formats specified at any time during the session.

Once the calling party has sent the offer it must be prepared to send and receive media in accordance with the offer session description.

When the called party receives the INVITE, it examines the offer session description and constructs a session description, which is the answer.

The answer must contain a media description line for each media description line in the offer, indicating whether it and the formats it specifies are accepted or not. If the media description as a whole is not accepted, the port number is set to zero. If it is accepted, the port number is set to the port that it wants to receive the media on. The formats that it accepts are reproduced in the media description of the answer and those that it rejects are removed. The order of the formats should remain the same, retaining the preferences of the party that sent the offer.

If a media description in the offer is followed by an attribute that indicates that it can only be received or sent (sendonly, recvonly), the corresponding media description in the answer must have this attribute reversed. If a media description in the offer has the inactive attribute, so should the corresponding media description in the answer. If the answer includes a bandwidth attribute for any media descriptor, this specifies the bandwidth that the called party would like the calling party to use when sending media—this need not be the same as in the offer.

Once the answer has been sent, perhaps as the payload of a SIP 183 Session Progress response or a 200 OK response, the called party must be prepared to send and receive media in accordance with the answer session description. When the calling party receives the answer, it can send media as specified in the answer session description.

If the calling party sends the offer of Figures 9.4 and 9.5, and the called party can support only GSM-AMR and baseline H.263, the session description for its answer will be similar to Figure 9.6.

The answer shown in Figure 9.6 has removed the audio and video formats that are not supported, and their associated attributes. The port numbers are different to those in the offer because they reflect the port numbers party B wishes to receive the media on. Since party B can only support video at a maximum rate of 64 kbit/s, the bandwidth attribute is modified.

```
v=0
o=partyB 2890852198 2890852198 IN IP6 partyB.mobiletelco.com
s=-
c=IN IP6 [ssss.tttt.uuuu.vvvv.wwww.xxxx.yyyy.zzzz]
b=CT:128
t=0 0
m=audio 46080 RTP/AVP 96
a=rtpmap:96 AMR/8000
m=video 47080 RTP/AVP 34
b=AS:64
a=rtpmap:34 H263/90000
```

Figure 9.6   Answer example.

When party A receives the answer, it can begin to send GSM-AMR and H.263 using RTP to the port numbers specified in the answer from party B. Party B can begin sending when it receives the SIP ACK for its 200 OK response.

At any point during the session, either participant can issue a new offer to modify characteristics of the session. Nearly all characteristics of a media stream can be modified, including the address, port or transport, the set of media formats, even changing media types. Probably most interesting is changing attributes, such as the frame rate and quality. This is analogous to the miscellaneous commands of H.245.

9.3   SIP in 3G Mobile Video Telephony

The use of SIP in 3G mobile networks is described in a number of 3GPP documents, including TS 24.228 "IP multimedia subsystem: Stage 2" and TS 24.229 "IP Multimedia Call Control Protocol based on Session Initiation Protocol (SIP) and Session Description Protocol (SDP): Stage 3". Aspects relating to video telephony are addressed in TS 26.235 "Packet switched conversational multimedia applications: Default Codecs (Release 5)" and TS 26.236 "Packet switched conversational multimedia applications; Transport Protocols (Release 5)." Many of these documents are still working drafts.

For SIP based video telephony the codecs that 3GPP specifies for audio are GSM-AMR as the mandatory audio codec and AMR-WB—a wideband CELP-based speech codec with a sampling rate of 16 kHz—twice that of other audio codecs we have discussed, as an optional codec. The mandatory video codec is baseline H.263, with H.263 Profile 3 Level 10 and MPEG4 SP@Level0 as optional codecs.

The establishment of a session differs from the simple example that we have described and makes use of UPDATE and PRACK methods. The

TABLE 9.2   Comparison of 3G-324M and SIP-Based Video Telephony

Function	3G-324M	3GPP SIP
Call establishment	Outside of scope of standard	SIP
Session establishment	H.245	SDP (within a SIP message)
Media transmission	H.223 MUX-PDUs over transparent circuit-switched channel	RTP, using UDP and IP, with RTCP for quality of service control
Control message transmission	H.223, with NSRP to provide acknowledged receipt	UDP, with acknowledgment at the application (SIP) layer
Video media types	H.263 baseline mandatory MPEG4 SP@L0 optional	H.263 mandatory H.263 Profile 3 optional MPEG4 SP@L0 optional
Audio media types	GSM-AMR mandatory G.723.1 optional	GSM-AMR mandatory AMR-WB optional

initial offer is sent in the first INVITE message. The first 183 Session Progress message that is sent back by the called terminal via one or more proxy servers contains the full capabilities of the called terminal. The calling terminal examines the answer and sends a PRACK for the received 183 Session Progress message, containing a revised offer if necessary, for example if the called terminal has more than one choice of codec for each media descriptor. If needed there are further opportunities to use SIP UPDATE and PRACK messages exchanged during session establishment to provide modified offers and responses to converge on an agreed set of media descriptions.

From the point of view of diagnosing problems SIP has the advantage over H.245 that the messages are text based and therefore easy to capture and review using readily available tools; however, this also makes them large compared to the compact binary representation provided by the PER encoded ASN.1 representations of H.245 messages. For this reason 3GPP specifies that over the air interface these messages should be compressed using a technique known as SigComp, defined in RFC 3320 "Signaling Compression (SigComp)" and RFC 3486 "Compressing the Session Initiation Protocol (SIP)."

Table 9.2 provides a comparison of 3G-324M and 3GPP SIP-based mobile video telephony.

9.4   Chapter Summary

The main focus of this book is on 3G-324M mobile video telephony. In later releases of 3G networks packet-based mobile video telephony is expected to become available, based on SIP and SDP.

In this chapter we have taken a look at how this works at a very high level. The SIP protocol is used to establish and control sessions. SDP, which may be carried in the payload of SIP messages, is used to describe them. The offer/answer approach is used to agree on the media that terminals will send to one another and what formats will be used. This is analogous to terminal capability exchange in H.245.

The specification of 3G systems based on SIP is not yet complete. The use of SIP signaling and other IETF protocols for packet video telephony initially appears simpler and easier to implement than 3G-324M. The IETF is working closely with 3GPP to identify areas where the relevant IETF standards need to be augmented to provide the full functionality needed in commercial mobile networks. This has seen a significant amount of activity and an expansion of the functionality required. This can be seen by comparing RFC 2543 "SIP: Session Initiation Protocol" dated March 1999, which is a document of 153 pages, with the 269 pages of RFC 3261 "SIP: Session Initiation Protocol" dated June 2002.

Chapter 10

Supplementary Services and Interworking

For 3G video telephony services to become popular mobile operators offering the service need to quickly establish a critical mass of users. There is a threshold at which the numbers of people with the service will tend to expand naturally—in the earlier stages of deployment the growth in users needs to be actively encouraged. The FOMA service based on 3G-324M grew at a rate well below what was forecast when it was initially launched by NTT DoCoMo but from late 2003 onward started to grow at well above forecasts. Enabling 3G video telephony customers to call people on Internet-based multimedia terminals and to call ISDN-based video telephone users will help to more quickly establish a population of users of video communication.

It is clear that whilst users appreciate the enhanced communication experience that is offered by video telephony, they will not accept any backward step in the supplementary services that are now available to them in 2G speech telephony.

In this chapter we will look at why interworking between different networks and standards is required, and how to provide it. We will also look at supplementary services. We will see that similar techniques to those used to provide interworking can enable rapid and cost-effective provision of these services.

10.1 Interworking between Different Networks

The development of standards has sought to promote open systems, where terminals and network equipment can be purchased from different vendors and be expected to work with each other. This approach has been followed for multimedia communications services as well as in many other areas of telecommunications and computer networks and has worked well. Unfortunately it is also true to say that these standards tend to be very network specific. H.320 has been adopted for ISDN-based video telephony and conferencing, H.323 and SIP for Internet-based multimedia, H.324 for GSTN, and 3G-324M for 3G circuit switched wireless networks.

The multimedia standards used in different networks have some things in common, but not enough to make interworking between terminals on different networks a simple matter that only involves some manipulation at the transport layer. The standards may differ in every aspect from the protocol used to establish the session, through the protocols used for multiplexing and transmitting multimedia information to the mandatory video and audio codecs specified.

10.1.1   Connecting to non-3G-324M multimedia terminals

There are a number of circumstances in which it is useful to be able to connect between 3G-324M mobile videophones and multimedia terminals on other networks using different multimedia protocols. Many businesses make use of H.323-based video telephony and conferencing and in Europe H.320-based videoconferencing over ISDN is popular within the business community. Enabling a 3G-324M terminal to join a conference based on either of these protocols allows participation by employees who are traveling and so would otherwise be unable to participate.

When 3G SIP multimedia terminals begin to become available, there is likely to be a sizeable population of users of 3G-324M video telephony. It will be very unfortunate if people with circuit-switched 3G-324M videophones are unable to make calls to people with SIP-based videophones. This provides a compelling case for providing the ability to interoperate with terminals using different multimedia protocols.

10.1.2   Connecting to speech terminals

In the specifications for Release 99 mobile video telephony mechanisms for a video call to fall back to voice if the called user is not able to support video capability are ill-defined at best. Currently available services do not offer this facility. This means that the caller has to know that the person they are calling is capable of receiving a video call. If the user makes a mistake and calls a terminal that is only capable of supporting speech the call will fail. Either the network will recognize that the call is between two incompatible terminals and refuse to establish a bearer, or the called terminal will produce a high pitched whine as it attempts to treat the information that it receives as speech.

Such unintended calls can be made because of user ignorance or error, but they can also occur when a user who is normally capable of making and receiving video calls moves into an area without 3G coverage. When mobile operators deploy 3G networks they are unlikely to make access available everywhere within their geographical area, particularly if they also operate an existing 2G network. 3G is likely to be available sooner in densely populated areas than in rural areas. For this reason handsets are capable of connecting over both 2G and 3G networks. A customer with video telephony capability may move into an area with only 2G coverage, or travel to a country that does not yet support video telephony services.

10.1.3 Multimedia gateways

To make it possible to call between multimedia terminals on different networks multimedia gateways are required. These convert between protocols and transcode between different media formats. A media gateway performs a similar function for speech-only terminals.

A media gateway is needed at the interface between the 3G mobile core network and the GSTN, to convert speech encoded in GSM-AMR to G.711 PCM speech at 64 kbit/s. To allow 3G-324M terminals to connect to multimedia terminals on different networks multimedia gateways are required that convert between 3G-324M information in 64 kbit/s PCM timeslots and other media, signaling, and network protocols.

An example of a multimedia gateway for converting between 3G-324M and SIP is shown in Figure 10.1. The system shown in the figure in fact includes an integrated *multimedia gateway* (MGW) and a *multimedia gateway controller* (MGC), but for now we will refer to this pair of components as a multimedia gateway.

The MGC is responsible for call signaling and session establishment. It uses this information to allocate resources in the MGW. The MGW is responsible for handling the media information. Its functions are to extract the media information from the incoming bit stream format, to transcode each media type from one format to another as necessary, and format it for transmission according to the requirements of the network and protocol that it is delivering the information to.

In the 3G-324M ↔ SIP example of Figure 10.1, the MGC coordinates establishment of a 3G-324M call on its circuit-switched side with establishment of a SIP-based call on its packet interface. To do this the MGC translates SS7 messages into SIP requests and responses and vice versa during the bearer establishment phase and once this phase is complete it allocates network and processing resources in the MGW to the call.

When the bearer is established on the circuit-switched side, the allocated MGW channel demultiplexes incoming H.223 MUX-PDUs to

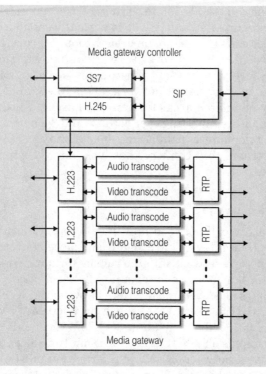

Figure 10.1 Structure of a 3G-324M ↔ SIP multimedia gateway.

extract H.245 messages and directs them to the MGC, which translates these into SIP messages that are forwarded to the SIP client. It takes incoming SIP messages, converts them into H.245 messages and forwards them to the MGW to be multiplexed and sent to the 3G-324M client. The MGC sees the capabilities being offered by both the SIP client and the 3G-324M client and uses its knowledge of the transcoding pairs that the MGW supports to establish compatible sets of capabilities for each side of the call. It uses these to complete session establishment on both sides.

Once the session has been established, media flows between the two endpoints via the MGW where it is transcoded as necessary. At no point in the call are the SIP and 3G-324M terminals aware of the fact that they are interacting with a gateway rather than another terminal.

The messages used by each terminal are very different; each set may include features and functionality that are not present in the other and the sequence in which comparable events take place in two protocols may be different. When the MGC translates between these sets of messages, this mapping can result in a slightly more complex sequence of messages being exchanged than would be the case in a direct terminal-to-terminal call using either protocol. An example of this can be seen by comparing the early stages of a SIP-to-SIP call with a 3G-324M originated call via a gateway to a SIP terminal.

As described in Chapter 9, in a SIP-to-SIP call the offer/answer model is used to set up a call. The originating terminal sends an INVITE message with an SDP description of the media it wishes to send and receive and the formats that the terminal supports. The called terminal sends an answer as a session description in a 183 Session progress message and, possibly after further offers and answers have been made, the terminals settle on media to be used for the call. When the originating terminal finally receives a 200 OK response from the called terminal and responds with an ACK, the session is fully established.

When a 3G-324M terminal originates a call to an SIP terminal the gateway gets an SS7 message asking to establish a bearer, and providing the address of the required destination. At this stage the MGC does not know the capabilities of the calling terminal. It can generate an INVITE message and forward it to the destination address, but it cannot include media descriptions, unless it makes the assumption that the calling terminal will at least support mandatory codecs and so include only these. The SIP side of the call must be established before the MGC accepts the incoming 3G-324M call and obtains the terminal capabilities of the calling terminal. Only then is it in a position to make a modified offer including any additional media formats supported.

If a call is originated in the opposite direction, a single incoming INVITE must be translated into multiple messages—SS7 messages to establish the bearer and the H.245 messages required for establishing a session.

The mapping of messages from one protocol to another may result in terminals being exercised by media gateways in ways that may not normally occur in a terminal to terminal call. This emphasizes the importance of terminals being fully standards compliant.

Commercial multimedia gateways will support multiple simultaneous calls and may support conversion between multiple protocols. Where large numbers of gateway ports or channels are required, it generally makes better economic sense to separate the MGW and MGC functions—a single MGC may be able to support many MGWs.

The underlying principles of a multimedia gateway are the same whichever pair of protocols it is converting between. A call between terminals on different networks using a gateway may result in longer call-establishment times than a call between two terminals using either standard. There may be increased delay in the transmission of media information and if transcoding is necessary a reduction in media quality is possible. These considerations mean that inserting more than one gateway in a call path is not advisable.

The DTG2000 from Dilithium Networks is one of the first multimedia gateways supporting 3G-324M to be commercially available. This uses patented transcoding techniques to reduce delay, maintain the highest possible media quality, and provide high channel density.

The traditional method of transcoding between two different media formats is known as tandem transcoding and involves fully decoding the incoming bitstream encoded in one format and reencoding it in another format. This is computationally intensive and introduces additional delay. Because the video and audio coding standards used in most multimedia systems of interest are based on similar underlying algorithms, as discussed in Chapter 3, the approach taken by Dilithium Networks is to transcode from one format to another without fully decoding. This approach dramatically reduces the computational resources required, resulting in higher capacity gateways. It also eliminates many of the causes of delay in tandem transcoding and provides media quality equivalent to, and in some cases better than, tandem transcoding.

10.2   Supplementary Services

For 3G video telephony to be successful, it is essential that at least the same range of supplementary services that are available to users of 2G mobile telephony services are provided, augmented where necessary to take account of the multimedia aspects of the service.

10.2.1   Mailbox services

Voice mailboxes are an established and standard feature of 2G mobile services. Providing the equivalent service for 3G video telephony implies enhancing the service to store and retrieve multimedia messages comprising both video and audio.

A way of interacting with the video mailbox is needed to allow the mailbox to be managed and messages to be retrieved, forwarded, or deleted by the user. H.245 provides the UserInputIndication message for this purpose but neither H.324 nor 3G-324M clearly state that support for this is mandatory, or even recommended. There is a danger that not all handsets will support this function.

10.2.2   Multipoint operation

Support for multiparty calling is likely to be an attractive feature in 3G video telephony services that will appeal to business users for conferencing and may also be popular with the youth market.

A *multipoint control unit* (MCU) is needed to provide the service. Speech-based multiparty conferencing is fairly straightforward; speech signals from each participant can simply be added by the MCU and retransmitted, perhaps with some amplitude adjustment. Video obviously cannot be treated in the same way. For continuous presence conferencing, where all participants see one another, the individual incoming video streams need to be combined to create a single output video stream that is transmitted back to all participants. If there are four participants, each operating at QCIF resolution, the video they are transmitting needs to be decoded and down sampled by the MCU to form four 88×72 pixel images, which are then reencoded as a single QCIF video stream for sending to all participants.

Alternative approaches include switching between participants using the highest amplitude speech signal to determine who is currently speaking, and transmit their video to all participants, or enabling manual control via an appointed chairperson.

MCUs that provide multipoint video conferencing are commercially available for H.323- and H.320-based multimedia services. There has been little standards-based activity in this area for H.324.

10.3 Supplementary Services and Interworking using Multimedia Gateways

Operators should consider their approach to implementing the infrastructure to provide supplementary services for 3G video telephony in the light of the expected longer-term evolution from circuit-switched 3G-324M to SIP-based video telephony. Providing two different sets of infrastructure, one for 3G-324M terminals and another for later SIP-based users could prove to be an expensive option. It would be far better if the same infrastructure could be used for both sets of users.

Fortunately, many of the components needed to provide supplementary services exist, at least in part, though they may not have the interfaces needed to use them with 3G-324M- or SIP-based terminals. An example that we have already mentioned is multipoint calling. This could be rapidly made available to 3G-324M video telephony customers by providing access to H.323-based MCUs via a 3G-324M ↔ H.323 multimedia gateway.

In a similar way, existing voicemail platforms with H.323- or SIP-based interfaces can be reused by adding video storage and a multimedia gateway that provides 3G-324M ↔ H.323 or 3G-324M ↔ SIP interworking. The 3G-324M ↔ SIP gateway can also be used to provide interworking between 3G-324M and SIP mobile video telephony customers.

Better support for interworking between speech-based handsets and 3G-324M handsets than the error-prone manual selection that is currently offered seems essential to improve usability. This can be done by using a gateway that can terminate a 3G-324M call and bridge across to a speech call. The gateway could handle the video aspect of one side of the call by declaring no capability to receive video in its terminal capability set, or it could support video and insert a video sequence locally at the gateway. This could be a visual indication that the remote user is currently unable to receive video.

It is clear that gateways and in particular multimedia gateways will be a significant part of 3G network infrastructure. Figure 10.2 shows a number of gateways placed within a 3G network and at the interfaces to other networks. Table 10.1 lists these gateways and their function.

Not all possible gateways are shown. Figure 10.2 illustrates how 3G-324M and SIP-based video telephony service can share infrastructure for supplementary services, such as video mail and multipoint platforms. It assumes that a video mail platform with a SIP interface is available. If

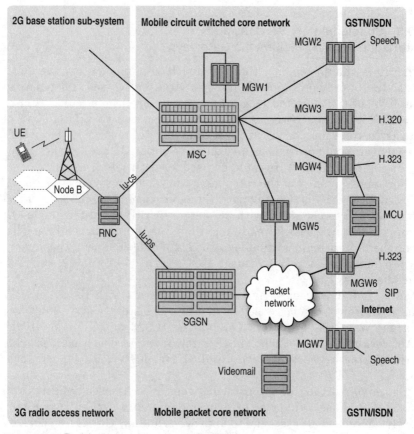

Figure 10.2 Positioning gateways in a 3G UMTS network.

TABLE 10.1   Function of Gateways in a 3G Network

Gateway	Multimedia	Between		Purpose
MGW1	Y	3G-324M	AMR	Interworking between 3G-324M and mobile speech terminals
MGW2	N	AMR	G.711	Interworking between mobile speech terminals and GSTN speech terminals
MGW3	Y	3G-324M	H.320	Interworking between 3G-324M and ISDN video conferencing
MGW4	Y	3G-324M	H.323	Inter-working between 3G-324M and H.323; provision of multipoint capability to 3G-324M users
MGW5	Y	3G-324M	SIP	Interworking between 3G-324M and SIP; provision of video mail services to 3G-324M users
MGW6	Y	SIP	H.323	Interworking between SIP and H.323; provision of multipoint capability to SIP users
MGW7	N	SIP	G.711	Interworking between SIP and GSTN speech terminals

instead a video mail platform with an H.323 interface is used, it could be placed behind MGW6, along with the MCU.

Although the SIP protocol has intrinsic support for multiparty calling, sending multiple video streams to each participant is very wasteful of bandwidth and requires that the user terminal can simultaneously decode and display multiple video streams. This is not necessary if the MCU functions are used via MGW6 in Figure 10.2.

The interworking capability illustrated in Figure 10.2 can be used by mobile network operators to provide service "bundles" to attract customers in different market sectors. For example a corporate package offering both 3G-324M 3G handsets and PC-based H.323 or SIP clients and cameras would appeal to businesses with a combination of office-based and traveling employees. The offer may be made more attractive if call forwarding is possible between the PC clients and the handsets, and vice versa.

10.4   Chapter Summary

In this chapter we have seen that supplementary services for mobile video telephony and access to multimedia terminals on other networks can be provided using multimedia gateways. Gateways with 3G-324M compliant interfaces may exercise handsets in ways that other handsets might not, emphasizing the need for thorough validation of the conformance of handsets to standards.

Multimedia gateways enable more rapid deployment of supplementary services. Interworking widens the base of users who can interact with each other and will help to get multimedia communication, and in particular video telephony, to the point where it will replace speech telephony as the de facto telephony service.

This is the last chapter of the book; I hope that you have learned what you expected to from it and that it has been successful in its aim of providing you with an understanding of 3G mobile video telephony.

Syntax of H.245 Messages Used in 3G-324M

This quick-reference guide is derived from Annex A of the H.245 Recommendation and is reproduced with the kind permission of ITU. It provides the syntax of H.245 messages referred to in the H.324 recommendation or in the 3GPP documentation and omits all H.245 messages that are not relevant to 3G-324M. It also omits H.245 codepoints related to the capability to open logical channels for data applications, encryption capability, multipoint capability and multilink capability. It removes OPTIONAL from those elements listed in H.245 as optional but which are mandatory in 3G-324M. Fixed values are provided (e.g., =TRUE or=FALSE where H.245 states BOOLEAN) where these are fixed in 3G-324M.

Codepoints related to mobile level 3 and the ALxM adaptation layers have been omitted, as have codepoints related to audio codecs other than G.723.1 and genericAudioCapability (for GSM-AMR), and video codecs other than H.263 and genericVideoCapability (for MPEG4). The codepoints included for H.263 relate only to baseline H.263 with Annexes D, E, F, and G and not to H.263 options.

This guide is meant to assist in understanding the syntax of messages associated with 3G-324M terminals; it is not intended to replace the H.245 recommendation, which should always be consulted for definitive syntax.

It is only suitable for the purpose intended; the complete syntax of H.245 is required by general purpose ASN.1 compilers that produce PER or BER messages.

A.1 Top Level Messages

```
MultimediaSystemControlMessage        ::=CHOICE
{
   request              RequestMessage,
   response             ResponseMessage,
   command              CommandMessage,
   indication           IndicationMessage,
}


RequestMessage        ::=CHOICE
{
   masterSlaveDetermination    MasterSlaveDetermination,
   terminalCapabilitySet       TerminalCapabilitySet,
   openLogicalChannel          OpenLogicalChannel,
   closeLogicalChannel         CloseLogicalChannel,
   requestChannelClose         RequestChannelClose,
   multiplexEntrySend          MultiplexEntrySend,
   requestMultiplexEntry       RequestMultiplexEntry,
   requestMode                 RequestMode,
   roundTripDelayRequest       RoundTripDelayRequest,
   maintenanceLoopRequest      MaintenanceLoopRequest,
}


ResponseMessage       ::=CHOICE
{
   masterSlaveDeterminationAck       MasterSlaveDeterminationAck,
   masterSlaveDeterminationReject    MasterSlaveDeterminationReject,
   terminalCapabilitySetAck          TerminalCapabilitySetAck,
   terminalCapabilitySetReject       TerminalCapabilitySetReject,
   openLogicalChannelAck             OpenLogicalChannelAck,
   openLogicalChannelReject          OpenLogicalChannelReject,
   closeLogicalChannelAck            CloseLogicalChannelAck,
   requestChannelCloseAck            RequestChannelCloseAck,
   requestChannelCloseReject         RequestChannelCloseReject,
   multiplexEntrySendAck             MultiplexEntrySendAck,
   multiplexEntrySendReject          MultiplexEntrySendReject,
   requestMultiplexEntryAck          RequestMultiplexEntryAck,
   requestMultiplexEntryReject       RequestMultiplexEntryReject,
   requestModeAck                    RequestModeAck,
   requestModeReject                 RequestModeReject,
   roundTripDelayResponse            RoundTripDelayResponse,
   maintenanceLoopAck                MaintenanceLoopAck,
   maintenanceLoopReject             MaintenanceLoopReject,
}


CommandMessage        ::=CHOICE
{
   maintenanceLoopOffCommand         MaintenanceLoopOffCommand,
   flowControlCommand                FlowControlCommand,
   endSessionCommand                 EndSessionCommand,
   miscellaneousCommand              MiscellaneousCommand,
   h223MultiplexReconfiguration      H223MultiplexReconfiguration,
}


IndicationMessage     ::=CHOICE
{
   functionNotUnderstood             FunctionNotUnderstood,
   masterSlaveDeterminationRelease   MasterSlaveDeterminationRelease,
   terminalCapabilitySetRelease      TerminalCapabilitySetRelease,
   openLogicalChannelConfirm         OpenLogicalChannelConfirm,
   requestChannelCloseRelease        RequestChannelCloseRelease,
```

```
   multiplexEntrySendRelease            MultiplexEntrySendRelease,
   requestMultiplexEntryRelease         RequestMultiplexEntryRelease,
   requestModeRelease                   RequestModeRelease,
   miscellaneousIndication              MiscellaneousIndication,
   h223SkewIndication                   H223SkewIndication,
   userInput                            UserInputIndication,
   vendorIdentification                 VendorIdentification,
   functionNotSupported                 FunctionNotSupported,
   flowControlIndication                FlowControlIndication,
}
--SequenceNumber is defined here as it is used in a number of
--Messages

SequenceNumber        ::=INTEGER (0..255)
```

A.2   Master/slave Determination

```
MasterSlaveDetermination       ::=SEQUENCE
{
   terminalType                INTEGER (0..255),         --default 128
   statusDeterminationNumber   INTEGER (0..16777215),    --random
}

MasterSlaveDeterminationAck    ::=SEQUENCE
{
   decision      CHOICE
   {
      master     NULL,
      slave      NULL
   },
}

MasterSlaveDeterminationReject          ::=SEQUENCE
{
   cause         CHOICE
   {
      identicalNumbers  NULL,
   },
}

MasterSlaveDeterminationRelease         ::=SEQUENCE
{
}
```

A.3   Capability Exchange Definitions

```
TerminalCapabilitySet        ::=SEQUENCE
{
   sequenceNumber             SequenceNumber,
   protocolIdentifier OBJECT IDENTIFIER,
   --set to the value {itu-t(0) recommendation(0) h(8) 245(0)
   --version(X)}, where X is the version in use by the terminal

   multiplexCapability        MultiplexCapability
   capabilityTable            SET SIZE(1..256)OF CapabilityTableEntry
   capabilityDescriptors      SET SIZE(1..256)OF CapabilityDescriptor
}
```

```
CapabilityTableEntry              ::=SEQUENCE
{
  capabilityTableEntryNumber            CapabilityTableEntryNumber,
  capability                            Capability
}
CapabilityDescriptor              ::=SEQUENCE
{
  capabilityDescriptorNumber            CapabilityDescriptorNumber,
  simultaneousCapabilities SET SIZE(1..256)OF AlternativeCapabilitySet
OPTIONAL
}

AlternativeCapabilitySet          ::=SEQUENCE     SIZE     (1..256)     OF
CapabilityTableEntryNumber

CapabilityTableEntryNumber        ::=INTEGER (1..65535)

CapabilityDescriptorNumber        ::=INTEGER (0..255)

TerminalCapabilitySetAck          ::=SEQUENCE
{
  sequenceNumber      SequenceNumber,
}

TerminalCapabilitySetReject           ::=SEQUENCE
{
  sequenceNumber      SequenceNumber,
  cause               CHOICE
{
    unspecified                         NULL,
    undefinedTableEntryUsed             NULL,
    descriptorCapacityExceeded          NULL,
    tableEntryCapacityExceeded          CHOICE
    {
      highestEntryNumberProcessed       CapabilityTableEntryNumber,
      noneProcessed                     NULL
    },
  },
}

TerminalCapabilitySetRelease          ::=SEQUENCE
{
}
```

A.3.1   Capability exchange definitions: Top level capability description

```
Capability              ::=CHOICE
{
  receiveVideoCapability                VideoCapability,
  transmitVideoCapability               VideoCapability,
  receiveAndTransmitVideoCapability     VideoCapability,

  receiveAudioCapability                AudioCapability,
  transmitAudioCapability               AudioCapability,
  receiveAndTransmitAudioCapability     AudioCapability,

  receiveUserInputCapability            UserInputCapability,
  transmitUserInputCapability           UserInputCapability,
  receiveAndTransmitUserInputCapability UserInputCapability,
}
```

A.3.2 Capability exchange definitions:
Multiplex capabilities

```
MultiplexCapability              ::=CHOICE
{
  h223Capability                 H223Capability,
}

H223Capability                   ::=SEQUENCE
{
  transportWithI-frames          = FALSE,
  videoWithAL1                   BOOLEAN,
  videoWithAL2                   BOOLEAN,
  videoWithAL3                   BOOLEAN,
  audioWithAL1                   BOOLEAN,
  audioWithAL2                   BOOLEAN,
  audioWithAL3                   BOOLEAN,
  maximumAl2SDUSize              INTEGER (0..65535),   --units octets
  maximumAl3SDUSize              INTEGER (0..65535),   --units octets
  maximumDelayJitter             INTEGER (0..1023),    --units millisecs
  h223MultiplexTableCapability   CHOICE
  {
    basic              NULL,
    enhanced           SEQUENCE
    {
      maximumNestingDepth               INTEGER (1..15),
      maximumElementListSize            INTEGER (2..255),
      maximumSubElementListSize         INTEGER (2..255),
    }
  },
  maxMUXPDUSizeCapability                BOOLEAN,
  nsrpSupport                            =TRUE,              --mandatory
  mobileOperationTransmitCapability      SEQUENCE
  {
      modeChangeCapability     BOOLEAN,
      h223AnnexA               = TRUE,         --mandatory
      h223AnnexADoubleFlag     BOOLEAN,
      h223AnnexB               = TRUE,         --mandatory
      h223AnnexBwithHeader     BOOLEAN,
  } OPTIONAL,
  bitRate        INTEGER (1..19200) --units of 100 bit/s
}
```

A.3.3 Capability exchange definitions:
Video capabilities

```
VideoCapability        ::=CHOICE
{
  h263VideoCapability        H263VideoCapability,
  genericVideoCapability     GenericCapability,       --used for MPEG4
}

H263VideoCapability    ::=SEQUENCE
{
  sqcifMPI           INTEGER (1..32)             --units 1/29.97 Hz
  qcifMPI            INTEGER (1..32)             --units 1/29.97 Hz
  cifMPI             INTEGER (1..32) OPTIONAL,   --units 1/29.97 Hz
  cif4MPI            INTEGER (1..32) OPTIONAL,   --units 1/29.97 Hz
  cif16MPI           INTEGER (1..32) OPTIONAL,   --units 1/29.97 Hz
  maxBitRate         INTEGER (1..19200),         --units 100 bit/s
```

```
    unrestrictedVector                          BOOLEAN,
    arithmeticCoding                            BOOLEAN,
    advancedPrediction                          BOOLEAN,
    pbFrames                                    BOOLEAN,
    temporalSpatialTradeOffCapability           BOOLEAN,
    errorCompensation                           BOOLEAN,
}
```

A.3.4   Capability exchange definitions: Audio capabilities

```
--For an H.223 multiplex, integers indicate the maximum number of audio
--frames per AL-SDU
AudioCapability        ::=CHOICE
{
  g7231                    SEQUENCE
  {
    maxAl-sduAudioFrames       INTEGER (1..256),
    silenceSuppression         BOOLEAN
  },
  genericAudioCapability     GenericCapability,
}
```

A.3.5   Capability exchange definitions: User input

```
UserInputCapability    ::= CHOICE
{
  basicString          NULL,   --characters 0-9, *, #
  iA5String            NULL,   --alphanumeric
  generalString        NULL,   --alphanumeric
  dtmf                 NULL,   --dtmf using signal and signalUpdate
}
```

A.3.6   Capability exchange definitions: Generic capability

```
GenericCapability      ::=SEQUENCE
{
  capabilityIdentifier       CapabilityIdentifier,
  maxBitRate                 INTEGER (0..4294967295) --Units 100 bit/s
  collapsing                 SEQUENCE OF GenericParameter  OPTIONAL,
  nonCollapsing              SEQUENCE OF GenericParameter  OPTIONAL,
  nonCollapsingRaw           OCTET STRING OPTIONAL,
  --Typically contains ASN.1 PER encoded data describing capability
}

CapabilityIdentifier    ::=CHOICE
{
  standard    OBJECT IDENTIFIER,
  --eg {itu-t(0) recommendation(0) h(8) generic caps(0) audio(1)amr(1)}
}

GenericParameter        ::=SEQUENCE
{
  parameterIdentifier        ParameterIdentifier,
  parameterValue             ParameterValue,
  supersedes                 SEQUENCE OF ParameterIdentifier OPTIONAL,
```

```
}

ParameterIdentifier    ::=CHOICE
{
  standard    INTEGER (0..127),   --Assigned by Capability specifications
}
ParameterValue      ::=CHOICE
{
  logical  NULL,  --Only acceptable if all entities include this option
  booleanArray  INTEGER (0..255),  --array of 8 logical types
  unsignedMin   INTEGER (0..65535),   --Look for min common value
  unsignedMax   INTEGER (0..65535),   --Look for max common value
  unsigned32Min  INTEGER (0..4294967295), --Look for min common value
  unsigned32Max  INTEGER (0..4294967295),--Look for max common value
  octetString   OCTET STRING, --non-collapsing octet string
  genericParameter  SEQUENCE OF GenericParameter,
  ...
}
```

A.4   Logical Channel Signaling Definitions

```
--"Forward" is used to refer to transmission in the direction from the
--terminal making the original request for a logical channel to the
--other terminal, and "reverse" is used to refer to the opposite
--direction of transmission, in the case of bi-directional channel
--requests.

OpenLogicalChannel    ::=SEQUENCE
{
  forwardLogicalChannelNumber          LogicalChannelNumber,
  forwardLogicalChannelParameters      SEQUENCE
  {
    dataType                DataType,
    multiplexParameters     CHOICE
    {
      h223LogicalChannelParameters   H223LogicalChannelParameters,
    },
    replacementFor    LogicalChannelNumber OPTIONAL
  },
  reverseLogicalChannelParameters      SEQUENCE
  {
    dataType                DataType,
    multiplexParameters            CHOICE
    {
      h223LogicalChannelParameters   H223LogicalChannelParameters,
    } OPTIONAL,
    ...,
    replacementFor            LogicalChannelNumber OPTIONAL
  } OPTIONAL,

  --Used to specify the reverse channel for bi-directional open
  --request. Not present for uni-directional channel request
}

LogicalChannelNumber  ::=INTEGER (1..65535)

DataType              ::=CHOICE
{
  nullData    NULL, --used when reverseLogicalChannel carries no media
```

```
   videoData    VideoCapability,
   audioData    AudioCapability,
}

H223LogicalChannelParameters  ::=SEQUENCE
{
 adaptationLayerType  CHOICE
 {
    al1Framed                        NULL,
    al1NotFramed                     NULL,
    al2WithoutSequenceNumbers        NULL,
    al2WithSequenceNumbers           NULL,
    al3                              SEQUENCE
    {
     controlFieldOctets    INTEGER (0..2),
     sendBufferSize        INTEGER (0..16777215)   — units octets
    },
 },
 segmentableFlag       BOOLEAN,
}


OpenLogicalChannelAck ::=SEQUENCE
{
  forwardLogicalChannelNumber        LogicalChannelNumber,
  reverseLogicalChannelParameters    SEQUENCE
  {
    reverseLogicalChannelNumber      LogicalChannelNumber,
    replacementFor                   LogicalChannelNumber OPTIONAL
  } OPTIONAL, --Not present for uni-directional channel request
}


OpenLogicalChannelReject       ::=SEQUENCE
{
  forwardLogicalChannelNumber        LogicalChannelNumber,
cause          CHOICE
  {
    unspecified                          NULL,
    unsuitableReverseParameters          NULL,
    dataTypeNotSupported                 NULL,
    dataTypeNotAvailable                 NULL,
    unknownDataType                      NULL,
    dataTypeALCombinationNotSupported    NULL,
    masterSlaveConflict                  NULL,
    replacementForRejected               NULL
  },
}


OpenLogicalChannelConfirm      ::=SEQUENCE
{
  forwardLogicalChannelNumber LogicalChannelNumber,
}


CloseLogicalChannel    ::=SEQUENCE
{
  forwardLogicalChannelNumber LogicalChannelNumber,
  source       CHOICE
  {
    user     NULL,
    lcse     NULL
  },
}
```

```
CloseLogicalChannelAck          ::=SEQUENCE
{
  forwardLogicalChannelNumber LogicalChannelNumber,
}

RequestChannelClose    ::=SEQUENCE
{
  forwardLogicalChannelNumber LogicalChannelNumber,
}

RequestChannelCloseAck          ::=SEQUENCE
{
  forwardLogicalChannelNumber LogicalChannelNumber,
}

RequestChannelCloseReject       ::=SEQUENCE
{
  forwardLogicalChannelNumber LogicalChannelNumber,
  cause               CHOICE
  {
    unspecified         NULL,
  },
}

RequestChannelCloseRelease      ::=SEQUENCE
{
  forwardLogicalChannelNumber LogicalChannelNumber,
}
```

A.5   H.223 Multiplex Table Definitions

```
MultiplexEntrySend      ::=SEQUENCE
{
  sequenceNumber        SequenceNumber,
  multiplexEntryDescriptors SET SIZE (1..15) OF MultiplexEntryDescriptor,
}

MultiplexEntryDescriptor        ::=SEQUENCE
{
  multiplexTableEntryNumber   MultiplexTableEntryNumber,
  elementList         SEQUENCE SIZE (1..256) OF MultiplexElement OPTIONAL
}

MultiplexElement        ::=SEQUENCE
{
  type                CHOICE
  {
    logicalChannelNumber        INTEGER(0..65535),
    subElementList              SEQUENCE SIZE (2..255) OF MultiplexElement
  },
  repeatCount CHOICE
  {
    finite                      INTEGER (1..65535), --repeats of type
    untilClosingFlag            NULL               --used for last element
  }
}

MultiplexTableEntryNumber       ::=INTEGER (1..15)
MultiplexEntrySendAck           ::=SEQUENCE
```

```
{
  sequenceNumber            SequenceNumber,
multiplexTableEntryNumber SET SIZE (1..15) OF MultiplexTableEntryNumber,
}


MultiplexEntrySendReject        ::=SEQUENCE
{
  sequenceNumber    SequenceNumber,
rejectionDescriptions SET SIZE (1..15) OF MultiplexEntryRejectionDescriptions,
}
MultiplexEntryRejectionDescriptions    ::=SEQUENCE
{
  multiplexTableEntryNumber    MultiplexTableEntryNumber,
  cause                CHOICE
  {
    unspecifiedCause            NULL,
    descriptorTooComplex        NULL,
  },
}


MultiplexEntrySendRelease        ::=SEQUENCE
{
multiplexTableEntryNumber SET SIZE (1..15) OF MultiplexTableEntryNumber,
}


RequestMultiplexEntry ::=SEQUENCE
{
entryNumbers  SET SIZE (1..15) OF MultiplexTableEntryNumber,
}


RequestMultiplexEntryAck        ::=SEQUENCE
{
  entryNumbers          SET SIZE (1..15) OF MultiplexTableEntryNumber,
}


RequestMultiplexEntryReject    ::=SEQUENCE
{
  entryNumbersSET SIZE (1..15) OF MultiplexTableEntryNumber,
  rejectionDescriptions        SET        SIZE        (1..15)            OF
RequestMultiplexEntryRejectionDescriptions,
}


RequestMultiplexEntryRejectionDescriptions    ::=SEQUENCE
{
  multiplexTableEntryNumber    MultiplexTableEntryNumber,
  cause                CHOICE
  {
  unspecifiedCause        NULL,
  },
}


RequestMultiplexEntryRelease    ::=SEQUENCE
{
  entryNumbers          SET SIZE (1..15) OF MultiplexTableEntryNumber,
}
```

A.6   Request Mode Definitions

```
--RequestMode is a list, in order or preference, of modes that a
--terminal would like to have transmitted to it.
RequestMode    ::=SEQUENCE
{
```

```
    sequenceNumber         SequenceNumber,
    requestedModes         SEQUENCE SIZE (1..256) OF ModeDescription,
}


RequestModeAck          ::=SEQUENCE
{
    sequenceNumber         SequenceNumber,
    response    CHOICE
    {
      willTransmitMostPreferredMode       NULL,
      willTransmitLessPreferredMode       NULL,
    },
}


RequestModeReject       ::=SEQUENCE
{
    sequenceNumber         SequenceNumber,
    cause                  CHOICE
    {
      modeUnavailable      NULL,
      requestDenied        NULL,
    },
}


RequestModeRelease      ::=SEQUENCE
{
}
```

A.6.1   Request mode definitions:
Mode description

```
ModeDescription         ::=SET SIZE (1..256) OF ModeElement

ModeElement    ::= SEQUENCE
{
    type                       ModeElementType,
    h223ModeParameters         H223ModeParameters OPTIONAL,
    logicalChannelNumber       LogicalChannelNumber OPTIONAL
}
ModeElementType         ::=CHOICE
{
    videoMode            VideoMode,
    audioMode            AudioMode,
}


H223ModeParameters      ::=SEQUENCE
{
    adaptationLayerType CHOICE
    {
      al1Framed                    NULL,
      al1NotFramed                 NULL,
      al2WithoutSequenceNumbers NULL,
      al2WithSequenceNumbers       NULL,
      al3
      {
        controlFieldOctets       INTEGER(0..2),
        sendBufferSize           INTEGER(0..16777215) --units octets
      },
    },
    segmentableFlag         BOOLEAN,
```

```
}
```

A.6.2   Request mode definitions:
Video modes

```
VideoMode                  ::=CHOICE
{
   h263VideoMode           H263VideoMode,
   genericVideoMode        GenericCapability        --used for MPEG4
}

H263VideoMode              ::=SEQUENCE
{
   resolution  CHOICE
   {
      sqcif       NULL,
      qcif        NULL,
      cif         NULL,         --unlikely
      cif4        NULL,         --very unlikely
      cif16       NULL,         --very unlikely
   },
   bitRate                     INTEGER (1..19200),      --units 100 bit/s
   unrestrictedVector          BOOLEAN,
   arithmeticCoding            BOOLEAN,
   advancedPrediction          BOOLEAN,
   pbFrames                    BOOLEAN,
}
```

A.6.3   Request mode definitions:
Audio modes

```
AudioMode       ::=CHOICE
{
   g7231                  CHOICE
   {
      noSilenceSuppressionLowRate       NULL,
      noSilenceSuppressionHighRate      NULL,
      silenceSuppressionLowRate         NULL,
      silenceSuppressionHighRate        NULL
   },
   genericAudioMode     GenericCapability,
}
```

A.7   Round Trip Delay Definitions

```
RoundTripDelayRequest            ::=SEQUENCE
{
   sequenceNumber        SequenceNumber,
}

RoundTripDelayResponse           ::=SEQUENCE
{
   sequenceNumber        SequenceNumber,
}
```

A.8   Maintenance Loop Definitions

```
MaintenanceLoopRequest        ::=SEQUENCE
{
  type          CHOICE
  {
    systemLoop                NULL,
    mediaLoop                 LogicalChannelNumber,
    logicalChannelLoop        LogicalChannelNumber,
  },
}

MaintenanceLoopAck     ::=SEQUENCE
{
  type          CHOICE
  {
    systemLoop                NULL,
    mediaLoop                 LogicalChannelNumber,
    logicalChannelLoop        LogicalChannelNumber,
  },
}

MaintenanceLoopReject         ::=SEQUENCE
{
  type          CHOICE
  {
    systemLoop                NULL,
    mediaLoop                 LogicalChannelNumber,
    logicalChannelLoop        LogicalChannelNumber,
  },
  cause                 CHOICE
  {
    canNotPerformLoop         NULL,
  },
}

MaintenanceLoopOffCommand     ::=SEQUENCE
{
}
```

A.9   Command Messages

A.9.1   Command message: Flow control

```
FlowControlCommand      ::=SEQUENCE
{
  scope         CHOICE
  {
    logicalChannelNumber      LogicalChannelNumber,
    wholeMultiplex            NULL
  },
  restriction CHOICE
  {
    maximumBitRate            INTEGER (0..16777215),    --units 100 bit/s
    noRestriction             NULL
  },
}
```

A.9.2   Command message: End session

```
EndSessionCommand        ::=CHOICE
{
  disconnect           NULL,
}
```

A.9.3   Command message: Miscellaneous commands

```
MiscellaneousCommand             ::=SEQUENCE
{
  logicalChannelNumber          LogicalChannelNumber,
  type        CHOICE
  {
    videoFreezePicture          NULL,
    videoFastUpdatePicture      NULL,
    videoFastUpdateGOB          SEQUENCE
    {
    firstGOB          INTEGER (0..17),
    numberOfGOBs      INTEGER (1..18)
  },
   videoTemporalSpatialTradeOff        INTEGER (0..31),
   videoFastUpdateMB                   SEQUENCE
   {
    firstGOB          INTEGER (0..255) OPTIONAL,
    firstMB           INTEGER (1..8192) OPTIONAL,
    numberOfMBs       INTEGER (1..8192),
   },
   maxH223MUXPDUsize             INTEGER(1..65535),       --units octets
  },
}
```

A.9.4   Command message: H.223 multiplex reconfiguration

```
H223MultiplexReconfiguration             ::=CHOICE
{
  h223ModeChange        CHOICE
  {
    toLevel0                        NULL,
    toLevel1                        NULL,
    toLevel2                        NULL,
    toLevel2withOptionalHeader      NULL,
  },
  h223AnnexADoubleFlag        CHOICE
  {
    start      NULL,
    stop       NULL,
  },
}
```

A.10   Indication Messages

A.10.1   Indication message: Function not supported

```
--This is used to return a complete request, response, or command that
--is not recognized
```

```
FunctionNotSupported  ::=SEQUENCE
{
  cause         CHOICE
  {
    syntaxError        NULL,
    semanticError      NULL,
    unknownFunction    NULL,
  },
  returnedFunction   OCTET STRING OPTIONAL,
}
```

A.10.2   Indication message: Miscellaneous indication

```
MiscellaneousIndication        ::=SEQUENCE
{
  logicalChannelNumber         LogicalChannelNumber,
  type                         CHOICE
  {
    videoTemporalSpatialTradeOff INTEGER (0..31),--current trade-off
    videoNotDecodedMBs                 SEQUENCE
    {
    firstMB                    INTEGER (1..8192),
    numberOfMBs                INTEGER (1..8192),
    temporalReference          INTEGER (0..255),
    },
  },
}
```

A.10.3   Indication message: H.223 logical channel skew

```
H223SkewIndication   ::=SEQUENCE
{
  logicalChannelNumber1        LogicalChannelNumber,
  logicalChannelNumber2        LogicalChannelNumber,
  skew                         INTEGER (0..4095),  --units millisecs
}
```

A.10.4   Indication message: Vendor identification

```
VendorIdentification   ::=SEQUENCE
{
  vendor        NonStandardIdentifier,
  productNumber OCTET STRING (SIZE(1..256)) OPTIONAL, --per vendor
  versionNumber OCTET STRING (SIZE(1..256)) OPTIONAL, --productNumber
}
```

A.10.5   Indication message: User input

```
UserInputIndication   ::=CHOICE
{
  alphanumeric                       GeneralString,
  userInputSupportIndication         CHOICE
  {
```

```
    nonStandard        NonStandardParameter,
    basicString        NULL,  --character 0-9, * and #
    iA5String          NULL,  --indicates iA5 string
    generalString      NULL,  --indicates general string
},
signal                        SEQUENCE
{
    signalType         IA5String (SIZE (1) ^ FROM ("0123456789#*ABCD!")),
    duration           INTEGER (1..65535) OPTIONAL,  --milliseconds
},
signalUpdate           SEQUENCE
{
    Duration           INTEGER (1..65535),  --milliseconds
},
}
```

A.10.6   Indication message: Flow control

```
FlowControlIndication ::=SEQUENCE
{
  scope                 CHOICE
  {
    logicalChannelNumber               LogicalChannelNumber,
    wholeMultiplex                     NULL
  },
  restriction CHOICE
  {
    maximumBitRate                     INTEGER (0..16777215),  --units 100 bit/s
    noRestriction                      NULL
  },
}
```

H.245 Signaling Entities

This appendix provides a description of the operation of each of the signaling entities used in the procedures of the ITU-T H.245 recommendation. It is an alternative and more concise way of describing them to that adopted in the recommendation (Annex C of H.245, which describes the procedures, is 100 pages long). It is meant to assist in understanding the procedures of H.245; it is not intended to replace the H.245 recommendation, which should always be consulted for definitive descriptions of the procedures and specifications of the functions of the signaling entities.

A set of tables is provided for each set of signaling entities. These show the states that it can be in, any internal state variables it has and any timers or counters it contains. These are followed by a list of the primitives that the SE user can use to instruct the signaling entity and a list of the primitives that the SE can use to inform the SE user of the current status of the procedure. This last list contains default parameter values. Any error primitives are then shown. The messages that the SE can send (and receive if it is not split into separate outgoing and incoming SEs) are then listed, and internal computation that the SE performs is shown.

Finally for each SE there is a table for each state that it can be in, showing what action it takes in response to received messages, primitives from its SE user or internal events such as timer expiry. When the action is that the SE uses a primitive to inform the user parameters are not shown unless they are different from the default values.

The word "send" is used in the table to indicate that a message is sent to the peer SE; "inform" indicates that the SE communicates a primitive to the SE user. When the SOURCE parameter in RELEASE.indication or REJECT.indication primitives is set to USER this means that the reason this primitive is being issued is because of an action by the remote SE user. Where it is set to PROTOCOL or the specific signaling entity (such as LCSE), this means that the SE has generated it as a result of a timer expiry or some other error condition.

Timers count down from an initial value and expire when they reach zero.

B.1   Master/Slave Determination Signaling Entity (MSDSE)

Allowed states of the MSDSE

State	Comment
IDLE	The state before and after the procedure is active
Incoming Awaiting Response (IAR)	The state when waiting for a final acknowledgment
Outgoing Awaiting Response (OAR)	The state immediately after sending a request

MSDSE internal state variables

Variable	Description	Comment
sv_STATUS	Terminal Status, one of the values {Master, Slave, Indeterminate}	Its value depends on the values of sv_TT and sv_SDNUM for the local and peer MSDSE
sv_TT	Terminal Type: (0...255)	Used in terminal status calculation
sv_SDNUM	Random number: $(0...2^{24})$	Used in terminal status calculation
sv_NCOUNT	Keeps count of the number of requests sent	—

MSDSE counters and timers

Identifier	Description	Comment
T106	Limits the time the SE will wait for a response	Should be slightly longer than the round-trip time
N100	Limits the number of attempts to establish terminal status	Usually set to 5

Primitives from SE user to MSDSE

Primitive	Purpose
DETERMINATION.request	Initiate master/slave determination

Primitives from MSDSE to SE user

Primitive	Purpose (and parameter default values)
DETERMINATION.confirm(TYPE)	Informs SE user of successful completion (MasterSlaveDeterminationAck.decision)
DETERMINATION.indication(TYPE)	Informs SE user of provisional agreement (sv_STATUS)
REJECT.indication	Informs SE user of failure

Error primitives from MSDSE to SE user

Error primitive	Error condition
ERROR.indication(A)	Local T106 timer expiry: no response from remote MSDSE
ERROR.indication(B)	Remote T106 timer expiry: remote has received no response
ERROR.indication(C)	Inappropriate MasterSlaveDetermination received
ERROR.indication(D)	Inappropriate MasterSlaveDeterminationReject received
ERROR.indication(E)	MasterSlaveDeterminationAck.decision inconsistent with sv_STATUS
ERROR.indication(F)	Maximum number of retries reached without success

Messages between MSDSE and peer MSDSE

Message	Type	Field values

MasterSlaveDetermination
{
    terminalType
    statusDeterminationNumber
}
	Request	terminalType = sv_TT
		statusDeterminationNumber =sv_SDNUM

MasterSlaveDeterminationAck
{
    decision
}
	Response	if sv_STATUS = master{
		    decision = slave
		}
		Else{
		    decision = master
		}

MasterSlaveDeterminationReject
{
    cause
}
	Response	cause = identicalNumbers

MasterSlaveDeterminationRelease	Command	—

Internal computations in MSDSE

Determine status.   The value of sv_STATUS is computed by comparing sv_TT for the local MSDSE with MasterSlaveDetermination.terminal Type received from the remote MSDSE. If these are identical, sv_SDNUM for the local MSDSE is compared with MasterSlave-Determination.statusDeterminationNumber received from the remote MSDSE.

Terminal type	Status determination number	Value of sv_status
=sv_TT	=sv_SDNUM	Indeterminate
=sv_TT	<sv_SDNUM	Master
=sv_TT	>sv_SDNUM	Slave
<sv_TT	Any value	Master
>sv_TT	Any value	Slave

Definition of MSDSE operation

The MSDSE is initialized with sv_TT assigned a Terminal Type value in the range (0...255) and sv_SDNUM assigned a random value in the range (0...16777215).

Current state=IDLE	
Message, event, or primitive	Actions
DETERMINATION.request	Set sv_NCOUNT = 1; Set timer T106; Send MasterSlaveDetermination; Move to state OAR

MasterSlaveDetermination	Determine status;
	If sv_STATUS=Indeterminate{
	    Send MasterSlaveDeterminationReject
	}
	Else{
	    Send MasterSlaveDeterminationAck;
	    Inform DETERMINATION.Indication;
	    Set timer 106;
	    Move to state IAR
	}

MasterSlaveDeterminationAck	Do nothing
MasterSlaveDeterminationReject	Do nothing
MasterSlaveDeterminationRelease	Do nothing

Current state=Outgoing Awaiting Response (OAR)	
Message, event, or primitive	Actions
MasterSlaveDeterminationAck	Reset timer 106 Send MasterSlaveDeterminationAck Inform DETERMINATION.confirm Move to state IDLE

MasterSlaveDeterminationReject	Reset timer 106;
	If sv_NCOUNT ≥ N100{
	　Inform ERROR.indication(F);
	　Inform REJECT.indication;
	　Move to state IDLE
	}
	Else{
	　Generate new value for sv_SDNUM;
	　Increment sv_NCOUNT;
	　Send MasterSlaveDetermination Request;
	　Set timer T106
	}

MasterSlaveDetermination	Reset timer 106
	Determine status
	If sv_STATUS=Indeterminate{
	　If sv_NCOUNT ≥ N100{
	　　Inform ERROR.indication(F);
	　　Inform REJECT.indication;
	　　Move to state IDLE
	　}
	　Else{
	　　Generate new value for sv_SDNUM;
	　　Increment sv_NCOUNT;
	　　Send MasterSlaveDetermination;
	　　Set timer T106
	　}
	}
	Else{
	　Send MasterSlaveDeterminationAck;
	　Inform DETERMINE.Indication;
	　Set timer 106;
	　Move to state IAR
	}

MasterSlaveDeterminationRelease	Reset timer 106; Inform ERROR.indication(B); Inform REJECT.indication; Move to state IDLE
Timer Expiry	Send MasterSlaveDeterminationRelease; Inform ERROR.indication(A); Move to state IDLE

Current state=Incoming awaiting response (IAR)	
Message, event, or primitive	Actions
MasterSlaveDetermination	Reset timer 106; Inform ERROR.indication(C); Inform REJECT.indication; Move to state IDLE

MasterSlaveDeterminationAck	Reset timer 106;
	If decision=sv_STATUS{
	   Inform DETERMINE.confirm
	}
	Else{
	   Inform ERROR.indication(E);
	   Inform REJECT.indication
	}
	Move to state IDLE

MasterSlaveDeterminationReject	Reset timer 106; Inform ERROR.indication(E); Inform REJECT.indication; Move to state IDLE
MasterSlaveDeterminationRelease	Reset timer 106 Inform ERROR.indication(B); Inform REJECT.indication Move to state IDLE
Timer Expiry	Inform ERROR.indication(A): Inform REJECT.indication; Move to state IDLE

B.2   Capability Exchange Signaling Entities (CESEs)

B.2.1   Outgoing CESE

Allowed states of the outgoing CESE

State	Comment
IDLE	The state before and after the procedure is active
Awaiting Response (AR)	The state immediately after sending a request

Outgoing CESE internal state variables

Variable	Description	Comment
out_SQ	Identifies the most recently sent request: (0...255)	Associates a response with a particular request

Outgoing CESE timers

Identifier	Description	Comment
T101	Limits the time the SE will wait for a response	Should be slightly longer than the round-trip time

Primitives from SE user to outgoing CESE

Primitive	Purpose
TRANSFER.request(PROTOID, MUXCAP, CAPTABLE, CAPDESCRIPTORS)	Initiates sending capabilities

Primitives from outgoing CESE t0 SE user

Primitive	Purpose (and parameter default values)
TRANSFER.confirm	Informs SE user of successful completion
REJECT.indication (SOURCE, CAUSE)	Informs SE user capability exchange refused (USER, null)

Messages sent by outgoing CESE

Message	Type	Field values

TerminalCapabilitySet{
  sequenceNumber
  protocolIdentifier
  multiplexCapability
  capabilityTable
  capabilityDescriptors
}
	Request	sequenceNumber = out_SQ
		protocolIdentifier = PROTID
		multiplexCapability = MUXCAP
		capabilityTable = CAPTABLE
		capabilityDescriptors = CAPDESCRIPTORS

TerminalCapabilitySetRelease	Command	None

Definition of outgoing CESE operation.   out_SEQ, is not initialized and can therefore be started at any random value.

Current state=IDLE	
Message, event, or primitive	Actions
TRANSFER.request	Increment out_SQ; Send TerminalCapabilitySet; Set timer T101; Move to state AR
TerminalCapabilitySetAck	Do nothing
TerminalCapabilitySetReject	Do nothing

Current state= Awaiting response (AR)	
Message, event, or primitive	Actions

TerminalCapabilitySetAck	If sequenceNumber=out_SQ{
	  Reset timer T101;
	  Inform TRANSFERconfirm;
	  Move to state IDLE
	}
	Else{
	  Do nothing
	}

TerminalCapabilitySetReject	If sequenceNumber=out_SQ{
	  Reset timer T101;
	  Inform REJECT.indication(USER,
	  TerminalCapabilitySetReject.cause);
	  Move to state IDLE
	}
	Else{
	  Do nothing
	}

Timer T101 Expiry	Send TerminalCapabilitySetRelease; Inform REJECT.indication(PROTOCOL, null); Move to state IDLE

B.2.2   Incoming CESE

Allowed states of the incoming CESE

State	Comment
Idle	The state before and after the procedure is active
Awaiting Response	The state immediately after receiving a request

Incoming CESE internal state variables

Variable	Description	Comment
in_SQ	Indicates the most recently received request:, (0...255)	Used to identify which request a response is associated with

Primitives from SE user to incoming CESE

Primitive	Purpose
TRANSFER.response	Initiates Ack: accepts capabilities
REJECT.request(CAUSE)	Initiates Reject: rejects capabilities

Primitives from incoming CESE to SE user

Primitive	Purpose (and parameter default values)
TRANSFER.indication (PROTOID, MUXCAP, CAPTABLE, CAPDESCRIPTORS)	Informs SE user of received capability set (TerminalCapabilitySet.protocolIdentifier TerminalCapabilitySet.multiplexCapability TerminalCapabilitySet.capabiltyTable TerminalCapabilitySet.capabilityDescriptors)
REJECT.indication(SOURCE)	Informs SE user of rejection (USER)

Messages sent by the incoming CESE

Message	Type	Field values
TerminalCapabilitySetAck{ sequenceNumber }	Response	sequenceNumber = in_SQ
TerminalCapabilitySetReject{ sequenceNumber cause }	Response	sequenceNumber = in_SQ cause = CAUSE

Definition of incoming CESE Operation

Current state=IDLE	
Message, event, or primitive	Actions
TerminalCapabilitySet	Set in_SQ=sequenceNumber; Inform TRANSFER.indication; Move to state AR
TerminalCapabilitySetRelease	Do nothing

Current state= Awaiting response (AR)	
Message, event, or primitive	Actions
TRANSFER.response	Send TerminalCapabilitySetAck; Move to state IDLE
REJECT.request	Send TerminalCapabilitySetReject; Move to state IDLE
TerminalCapabilitySet	Set in_SQ=sequenceNumber; Inform REJECT.indication; Inform TRANSFER.indication
TerminalCapabilitySetRelease	Inform REJECT.indication(PROTOCOL); Move to state IDLE

B.3   Logical Channel Signaling Entities (LCSEs)

B.3.1   Outgoing LCSE

Allowed states of the outgoing LCSE

State	Comment
Released (REL)	The state when the Logical Channel is not open
Awaiting Establishment (AE)	The state immediately after sending an OpenLogicalChannel request
Established (EST)	The state when the Logical Channel is open
Awaiting Release (AR)	The state immediately after sending an CloseLogicalChannel request

Outgoing LCSE internal state variables

Variable	Description	Comment
out_LCN	Indicates the Logical Channel Number: (0...255)	Identifies the outgoing LCSEs

Outgoing LCSE timers

Identifier	Description	Comment
T103	Limits the time the SE will wait for a response	Should be slightly longer than the round-trip time.

Primitives from SE user to outgoing LCSE

Primitive	Purpose
ESTABLISH.request(FORWARD_PARAM)	Initiates opening logical channel
RELEASE.request	Initiates closing logical channel

Primitives from outgoing LCSE to SE user

Primitive	Purpose (and parameter default values)
ESTABLISH.confirm	Informs SE user open logical channel is successful
RELEASE.confirm	Informs SE user close logical channel is successful
RELEASE.indication (SOURCE, CAUSE)	Informs SE user logical channel being closed (CloseLogicalChannel.source, null)

Error primitives from outgoing LCSE to SE user

Error primitive	Error condition
ERROR.indication(A)	Inappropriate OpenLogicalChannelAck received
ERROR.indication(B)	Inappropriate OpenLogicalChannelReject received
ERROR.indication(C)	Inappropriate CloseLogicalChannelAck received
ERROR.indication(D)	T103 timer expiry: no response from remote incoming LCSE

Messages sent by outgoing LCSE

Message	Type	Default values

OpenLogicalChannel{
  forwardLogicalChannelNumber
  forwardLogicalChannelParameters
}
	Request	forwardLogicalChannelNumber
		= out_LCN
		forwardLogicalChannelParameters
		= FORWARD_PARAM

CloseLogicalChannel{
  forwardLogicalChannelNumber
  source
}
	Request	forwardLogicalChannelNumber
		= out_LCN
		source = user

Definition of outgoing LCSE operation.   out_LCN is initialized to the number of the logical channel it is associated with.

Current state=Released (REL)	
Message, event, or primitive	Actions
ESTABLISH.request	Send OpenLogicalChannel; Set timer T103; Move to state AE
OpenLogicalChannelAck	Inform ERROR.indication(A)
OpenLogicalChannelReject	Inform ERROR.indication(B)
CloseLogicalChannelAck	Do nothing

Current state=Awaiting Establishment (AE)	
Message, event, or primitive	Actions
OpenLogicalChannelAck	Reset timer T103; Inform ESTABLISH.confirm; Move to state EST
OpenLogicalChannelReject	Reset timer T103; Inform RELEASE.indication(USER, OpenLogicalChannelReject.cause); Move to state REL
CloseLogicalChannelAck	Do nothing
RELEASE.request	Reset timer T103; Send CloseLogicalChannel Request; Set Ttimer 103; Move to state AR
Timer Expiry	Reset timer T103; Inform ERROR.indication(D); Send CloseLogicalChannel Request; Inform RELEASE.indication(LCSE, null); Move to state REL

Current state=Established (EST)	
Message, event, or primitive	Actions
RELEASE.request	Send CloseLogicalChannel Request; Set timer T103; Move to state AR
OpenLogicalChannelAck	Do nothing
OpenLogicalChannelReject	Inform ERROR.indication(B); Inform RELEASE.indication(LCSE, null); Move to state REL
CloseLogicalChannelAck	Inform ERROR.indication(C); Inform RELEASE.indication(LCSE, null); Move to state REL

Current state=Awaiting Release (AR)	
Message, event, or primitive	Actions
CloseLogicalChannelAck	Reset timer T103; Inform RELEASE.confirm; Move to state REL
CloseLogicalChannelReject	Reset timer T103; Inform RELEASE.confirm; Move to state REL
OpenLogicalChannelAck	Do nothing
Timer Expiry	Inform ERROR.indication(D); Inform RELEASE.confirm; Move to state REL
ESTABLISH.request	Reset timer T103; Send OpenLogicalChannel Request; Set timer T103; Move to state AE

B.3.2   Incoming LCSE

Allowed states of the incoming LCSE

State	Comment
Released (REL)	The state when the Logical Channel is not open
Awaiting Establishment (AE)	The state after receiving an OpenLogicalChannel request
Established (EST)	The state when the Logical Channel is open

Incoming LCSE internal state variables

Variable	Description	Comment
in_LCN	Indicates the Logical Channel Number: (0...255)	Used to identify the incoming LCSE

Primitives from SE user to incoming LCSE

Primitive	Purpose
ESTABLISH.response	Initiates Ack: accepts open logical channel
RELEASE.request(CAUSE)	Initiates Reject: rejects open logical channel

Primitives from incoming LCSE to SE user

Primitive	Purpose (and parameter default values)
ESTABLISH.indication (FORWARD_PARAM)	Informs SE user of logical channel request (OpenLogicalChannel.forwardLogicalChannelParameters)
RELEASE.indication (SOURCE)	Informs SE user logical channel is closing (CloseLogicalChannel.source)

Messages sent by incoming LCSE

Message	Type	Field

OpenLogicalChannelAck{
   forwardLogicalChannelNumber
 }
	Response	forwardLogicalChannelNumber = in_LCN

OpenLogicalChannelReject{
   forwardLogicalChannelNumber
   cause
 }
	Response	forwardLogicalChannelNumber = in_LCN
		 cause = CAUSE

CloseLogicalChannelAck{
   forwardLogicalChannelNumber
 }
	Response	forwardLogicalChannelNumber = in_LCN

Definition of incoming LCSE operation. in_LCN is initialized to the number of the logical channel it is associated with.

Current state=Released (REL)	
Message, event, or primitive	Actions
OpenLogicalChannel	Inform ESTABLISH.indication; Move to state AE
CloseLogicalChannel	Send CloseLogicalChannelAck

Current state=Awaiting Establishment (AE)	
Message, event, or primitive	Actions
ESTABLISH.response	Send OpenLogicalChannelAck; Move to state EST
RELEASE.request	Send OpenLogicalChannelReject; Move to state REL
CloseLogicalChannel	Send CloseLogicalChannelAck; Inform RELEASE.indication; Move to state REL
OpenLogicalChannel	Inform RELEASE.indication(USER); Inform ESTABLISH.indication; Move to state AE

Current state=Established (EST)	
Message, event, or primitive	Actions
CloseLogicalChannel	Send CloseLogicalChannelAck; Inform RELEASE.indication; Move to state REL
OpenLogicalChannel	Inform RELEASE.indication(USER); Inform ESTABLISH.indication; Move to state AE

B.4   Bidirectional Logical Channel Signaling Entities (B-LCSEs)

B.4.1   Outgoing B-LCSE

Allowed states of the outgoing B-LCSE

State	Comment
Released (REL)	The state when the Logical Channel is not open
Awaiting Establishment (AE)	The state immediately after sending an OpenLogicalChannel request
Established (EST)	The state when the Logical Channel is open
Awaiting Release (AR)	The state immediately after sending an CloseLogicalChannel request

Outgoing B-LCSE Internal state variables

Variable	Description	Comment
out_LCN	Indicates the Logical Channel Number: (0...255)	Identifies the outgoing B-LCSE

Outgoing B-LCSE timers

Identifier	Description	Comment
T103	Limits the time the SE will wait for a response	Should be slightly longer than the round-trip time.

Primitives from SE user to outgoing B-LCSE

Primitive	Purpose
ESTABLISH.request(FORWARD_PARAM, REVERSE_PARAM)	Initiates opening logical channel
RELEASE.request	Initiates closing logical channel

Primitives from outgoing B-LCSE to SE user

Primitive	Purpose (and parameter default values)
ESTABLISH.confirm (REVERSE_DATA)	Informs SE user open logical channel is successful (OpenLogicalChannelAck. reverseLogicalChannelParameters)
RELEASE.confirm	Informs SE user close logical channel is successful
RELEASE.indication (SOURCE, CAUSE)	Informs SE user logical channel is closing (CloseLogicalChannel.source, null)

Error primitives from outgoing B-LCSE to SE user

Error primitive	Error condition
ERROR.indication(A)	Inappropriate OpenLogicalChannelAck message received
ERROR.indication(B)	Inappropriate OpenLogicalChannelReject message received
ERROR.indication(C)	Inappropriate CloseLogicalChannelAck message received
ERROR.indication(D)	T103 timer expiry: no response from remote incoming B-LCSE

Messages sent by outgoing B-LCSE

Message	Type	Default field values
OpenLogicalChannel{ forwardLogicalChannelNumber forwardLogicalChannelParameters reverseLogicalChannelParameters }	Request	forwardLogicalChannelNumber = out_LCN forwardLogicalChannelParameters = FORWARD_PARAM reverseLogicalChannelParameters = REVERSE_PARAM
CloseLogicalChannel{ forwardLogicalChannelNumber source }	Request	forwardLogicalChannelNumber = out_LCN source=user
OpenLogicalChannelConfirm{ forwardLogicalChannelNumber }	Indication	forwardLogicalChannelNumber = out_LCN

Definition of outgoing B-LCSE operation. out_LCN is initialized to the number of the bidirectional logical channel it is associated with.

Current state=Released (REL)	
Message, event, or primitive	Actions
ESTABLISH.request	Send OpenLogicalChannel; Set timer T103; Move to state AE
OpenLogicalChannelAck	Inform ERROR.indication(A)
OpenLogicalChannelReject	Inform ERROR.indication(B)
CloseLogicalChannelAck	Do nothing

Current state=Awaiting Establishment (AE)	
Message, event, or primitive	Actions
OpenLogicalChannelAck	Reset timer T103; Inform ESTABLISH.confirm; Send OpenLogicalChannelConfirm; Move to state EST
OpenLogicalChannelReject	Reset timer T103; Inform RELEASE.indication(USER, OpenLogicalChannelReject.cause); Move to state REL
CloseLogicalChannelAck	Do nothing
RELEASE.request	Reset timer T103; Send CloseLogicalChannel Request; Set timer 103; Move to state AR
Timer Expiry	Reset timer T103; Inform ERROR.indication(D); Send CloseLogicalChannel Request; Inform RELEASE.indication(B-LCSE, null); Move to state REL

Current state=Established (EST)	
Message, event, or primitive	Actions
RELEASE.request	Send CloseLogicalChannel Request; Set timer T103; Move to state AR
OpenLogicalChannelAck	Do nothing
OpenLogicalChannelReject	Inform ERROR.indication(B); Inform RELEASE.indication(B-LCSE, null); Move to state REL
CloseLogicalChannelAck	Inform ERROR.indication(C); Inform RELEASE.indication(B-LCSE, null); Move to state REL

Current state=Awaiting Release (AR)	
Message, event, or primitive	Actions
CloseLogicalChannelAck	Reset timer T103; Inform RELEASE.confirm; Move to state REL
CloseLogicalChannelReject	Reset timer T103; Inform RELEASE.confirm; Move to state REL
OpenLogicalChannelAck	Do nothing
Timer Expiry	Inform ERROR.indication(D); Inform RELEASE.confirm; Move to state REL
ESTABLISH.request	Reset timer T103; Send OpenLogicalChannel Request; Set timer T103; Move to state AE

B.4.2 Incoming B-LCSE

Allowed states of the incoming B-LCSE

State	Comment
Released (REL)	The state when the Logical Channel is not open
Awaiting Establishment (AE)	The state immediately after receiving an OpenLogicalChannel request
Awaiting Confirmation (AC)	The state immediately after sending an OpenLogicalChannelAck response
Established (EST)	The state when the Logical Channel is open

Incoming B-LCSE internal state variables

Variable	Description	Comment
in_LCN	Indicates the Logical Channel Number: (0...255)	Identifies the incoming LCSE

Incoming B-LCSE timers

Identifier	Description	Comment
T103	Limits the time the SE will wait for a Confirm indication	Should be slightly longer than the round-trip time

Primitives from SE user to incoming B-LCSE

Primitive	Purpose
ESTABLISH.response (REVERSE_DATA)	Initiates Ack: accepts open logical channel
RELEASE.request (CAUSE)	Initiates Reject: rejects open logical channel

Primitives from incoming B-LCSE to SE user

Primitive	Purpose (and parameter default values)
ESTABLISH.indication (FORWARD_PARAM, REVERSE_PARAM)	Informs SE user of logical channel request (OpenLogicalChannel.forwardLogicalChannelParameters, OpenLogicalChannel.reverseLogicalChannelParameters)
ESTABLISH.confirm	Informs SE user open logical channel is successful
RELEASE.indication (SOURCE)	Informs SE user logical channel is closing (CloseLogicalChannel.source)

Error primitives from incoming B-LCSE to SE user

Error primitive	Error condition
ERROR.indication(E)	Inappropriate OpenLogicalChannelConfirm received
ERROR.indication(F)	T103 timer expiry: no response from remote outgoing B-LCSE

Messages sent by incoming B-LCSE

Message	Type	Field

OpenLogicalChannelAck{
forwardLogicalChannelNumber
reverseLogicalChannelParameters
}
	Response	forwardLogicalChannelNumber
		= in_LCN
		reverseLogicalChannelParameters
		=REVERSE_PARAM

OpenLogicalChannelReject{
forwardLogicalChannelNumber
cause
}
	Response	forwardLogicalChannelNumber
		= in_LCN
		cause = CAUSE

CloseLogicalChannelAck{
forwardLogicalChannelNumber
}
	Response	forwardLogicalChannelNumber
		= in_LCN

Definition of incoming B-LCSE operation.   in_LCN is initialized to the number of the bidirectional logical channel it is associated with.

Current state=Released (REL)	
Message, event, or primitive	Actions
OpenLogicalChannel	Inform ESTABLISH.indication; Move to state AE
CloseLogicalChannel	Send CloseLogicalChannelAck
OpenLogicalChannelConfirm	Do nothing

Current state=Awaiting Establishment (AE)	
Message, event, or primitive	Actions
ESTABLISH.response	Send OpenLogicalChannelAck; Set timer T103; Move to state AC
RELEASE.request	Send OpenLogicalChannelReject; Move to state REL
CloseLogicalChannel	Send CloseLogicalChannelAck; Inform RELEASE.indication; Move to state REL
OpenLogicalChannel	Inform RELEASE.indication(USER); Inform ESTABLISH.indication;
OpenLogicalChannelConfirm	Inform ERROR.indication(E); Inform RELEASE.indication(B-LSCE); Move to state REL

| Current state=Awaiting Confirmation (AC) ||
Message, event, or primitive	Actions
Timer Expiry	Inform ERROR.indication(F); Inform RELEASE.indication; Move to state REL
OpenLogicalChannelConfirm	Reset timer T103; Inform ESTABLISH.confirm; Move to state EST
CloseLogicalChannel	Reset timer T103; Send CloseLogicalChannelAck; Inform RELEASE.indication; Move to state REL
OpenLogicalChannel	Reset timer T103; Inform RELEASE.indication(USER); Inform ESTABLISH.indication; Move to state AE

| Current state=Established (EST) ||
Message, event, or primitive	Actions
CloseLogicalChannel	Send CloseLogicalChannelAck; Inform RELEASE.indication; Move to state REL
OpenLogicalChannel	Inform RELEASE.indication(USER); Inform ESTABLISH.indication; Move to state AE
OpenLogicalChannelConfirm	Do nothing

B.5 Close Logical Channel Signaling Entities (CLCSEs)

B.5.1 Outgoing CLCSE

Allowed states of the outgoing CLCSE

State	Comment
IDLE	The state before and after the procedure is active
Awaiting Release (AR)	The state immediately after sending an RequestChannelClose request

Outgoing CLCSE internal state variables

Variable	Description	Comment
out_LCN	Indicates the Logical Channel Number: (0...255)	Identifies the outgoing CLCSE

Outgoing CLCSE timers

Identifier	Description	Comment
T108	Limits the time the SE will wait for a response	Should be slightly longer than the round-trip time.

Primitives from SE user to outgoing CLCSE

Primitive	Purpose
CLOSE.request	Initiates request close channel

Primitives from outgoing CLCSE to SE user

Primitive	Purpose (and parameter default values)
CLOSE.confirm	Informs SE user of success of the procedure
REJECT.indication (SOURCE, CAUSE)	Informs SE user of rejection (USER, null)

Messages sent by outgoing CLCSE

Message	Type	Default values
RequestChannelClose{ forwardLogicalChannelNumber }	Request	forwardLogicalChannelNumber = out_LCN
RequestChannelCloseRelease{ forwardLogicalChannelNumber }	Indication	forwardLogicalChannelNumber = out_LCN

Definition of outgoing CLCSE operation. out_LCN is initialized to the number of the logical channel it is associated with.

Current state=IDLE	
Message, event, or primitive	Actions
CLOSE.request	Send RequestChannelClose; Set timer T108; Move to state AR
RequestChannelCloseAck	Do nothing
RequestChannelCloseReject	Do nothing

Current state=Awaiting Release (AR)	
Message, event, or primitive	Actions
RequestChannelCloseAck	Reset timer T108; Inform CLOSE.confirm; Move to state IDLE
RequestChannelCloseReject	Reset timer T108; Inform REJECT.indication (USER, RequestChannelCloseReject.cause); Move to state IDLE
Timer Expiry	Send RequestChannelCloseRelease; Inform REJECT.indication(PROTOCOL, null); Move to state IDLE

B.5.2   Incoming CLCSE

Allowed states of the incoming CLCSE

State	Comment
IDLE	The state before and after the procedure has executed.
Awaiting Release (AR)	The state immediately after receiving a request

Incoming CLCSE internal state variables

Variable	Description	Comment
in_LCN	Indicates the Logical Channel Number: (0...255)	Identifies the incoming CLCSE

Primitives from SE user to incoming CLCSE

Primitive	Purpose
CLOSE.response	Initiates Ack: accepts close channel request
REJECT.request(CAUSE)	Initiates Reject: rejects close channel request

Primitives from incoming CLCSE to SE user

Primitive	Purpose (and parameter default values)
CLOSE.indication	Informs SE user of close channel request
REJECT.indication (SOURCE)	Informs SE user of rejection (USER)

Messages sent by incoming CLCSE

Message	Type	Field
RequestCloseChannelAck{ forwardLogicalChannelNumber }	Response	forwardLogicalChannelNumber = in_LCN
RequestCloseChannelReject{ forwardLogicalChannelNumber cause }	Response	forwardLogicalChannelNumber = in_LCN cause = CAUSE

Definition of incoming CLCSE operation. in_LCN is initialized to the number of the logical channel it is associated with.

Current state=IDLE	
Message, event, or primitive	Actions
RequestChannelClose	Inform CLOSE.indication; Move to state AR
RequestChannelCloseRelease	Do nothing

Current state=Awaiting Response (AR)	
Message, event, or primitive	Actions
CLOSE.response	Send CloseLogicalChannelAck; Move to state IDLE
REJECT.request	Send RequestChannelCloseReject; Move to state IDLE
RequestChannelCloseRelease	Inform REJECT.indication(PROTOCOL); Move to state IDLE
RequestChannelClose	Inform REJECT.indication; Inform CLOSE.indication;

B.6   Multiplex Table Signaling Entities (MTSEs)

B.6.1   Outgoing MTSE

Allowed states of the outgoing MTSE

State	Comment
IDLE	The state before and after the procedure has executed
Awaiting Response (AR)	The state immediately after sending a request

Outgoing MTSE internal state variables

Variable	Description	Comment
out_SQ	Indicates the most recently sent request: (0…255)	Associates a response with a particular request
out_ENUM	Indicates the outgoing MTSE instance: (1…15)	Identifies the outgoing MTSE and related multiplexTableEntryNumber

Outgoing MTSE timers

Identifier	Description	Comment
T104	Limits the time the SE will wait for a response	Should be slightly longer than the round-trip time

Primitives from SE user to outgoing MTSE

Primitive	Purpose
TRANSFER.request(MUX-DESCRIPTOR)	Initiates multiplex entry send

Primitives from outgoing MTSE to SE user

Primitive	Purpose (and parameter default values)
TRANSFER.confirm	Informs SE user of the success of the procedure
REJECT.indication (SOURCE, CAUSE)	Informs SE user of rejection (USER, null)

Messages sent by outgoing MTSE

Message	Type	Default values
MultiplexEntrySend{ sequenceNumber multiplexEntryDescriptors. multiplexTableEntryNumber multiplexEntryDescriptors. elementlist }	Request	sequenceNumber = out_SQ multiplexEntryDescriptors. multiplexTableEntryNumber = out_ENUM multiplexEntryDescriptors.elementlist = MUX_DESCRIPTOR
MultiplexEntrySendRelease{ multiplexTableEntryNumber }	Indication	multiplexTableEntryNumber = out_ENUM

Definition of outgoing MTSE operation. The Outgoing MTSE is initialized with the value of out_ENUM corresponding to the multiplex-TableEntryNumber that the outgoing MTSE refers to. The value of out_SQ is not initialized, sohence may start at any value.

Current state=IDLE	
Message, event, or primitive	Actions
TRANSFER.request	Increment out_SQ; Send MultiplexEntrySend Request; Set timer T104; Move to state AR
MultiplexEntrySendAck	Do nothing
MultiplexEntrySendReject	Do nothing

Current state=Awaiting Response (AR)	
Message, event, or primitive	Actions

MultiplexEntrySendAck	If sequenceNumber=out_SQ{
	  Reset timer T104;
	  Inform TRANSFERconfirm;
	  Move to state IDLE
	}
	Else{
	  Do nothing
	}

TerminalCapabilitySetReject	If sequenceNumber=out_SQ{
	  Reset timer T104;
	  Inform REJECT.indication(USER,
	MultiplexEntrySendReject.cause);
	  Move to state IDLE
	}
	Else{
	  Do nothing
	}

TRANSFER.request	Reset timer T104; Increment out_SQ; Send MultiplexEntrySend; Set timer T104
Timer Expiry	Send MultiplexEntrySendRelease; Inform REJECT.indication(PROTOCOL); Move to state IDLE

B.6.2   Incoming MTSE

Allowed states of the incoming MTSE

State	Comment
IDLE	The state before and after the procedure has executed
Awaiting Response (AR)	The state immediately after receiving a request

Incoming MTSE internal state variables

Variable	Description	Comment
in_SQ	Indicates the most recently received request: (0...255)	Used to identify which request a response is associated with
In_ENUM	Indicates the incoming MTSE instance: (1...15)	Used to identify incoming MTSE and related multiplexTableEntryNumber

Primitives from SE user to incoming MTSE

Primitive	Purpose
TRANSFER.response	Initiates Ack: accepts multiplex table entries
REJECT.request(CAUSE)	Initiates Reject: rejects multiplex table entries

Primitives from incoming MTSE to SE user

Primitive	Default parameter values
TRANSFER.indication (MUX-DESCRIPTOR)	Informs SE user of MultiplexEntrySend request (MultiplexEntrySend.multiplexEntryDescriptors. elementList)
REJECT.indication(SOURCE)	Informs SE user of rejection (USER)

Messages sent by the incoming MTSE

Message	Type	Default value
MultiplexEntrySendAck{ SequenceNumber MultiplexTableEntryNumber }	Response	sequenceNumber = in_SQ MultiplexTableEntryNumber = in_ENUM
MultiplexEntrySendReject{ SequenceNumber rejectionDescriptions. MultiplexTableEntryNumber rejectionDescriptions cause }	Response	sequenceNumber = in_SQ rejectionDescriptions. MultiplexTableEntryNumber = in_ENUM rejectionDescriptions cause = CAUSE

Definition of incoming MTSE operation. The Incoming MTSE is initialized with the value in_ENUM corresponding to the multiplexTable-EntryNumber that the incoming MTSE refers to.

Current state=IDLE	
Message, event, or primitive	Actions
MultiplexEntrySend	Set in_SQ=sequenceNumber; Inform TRANSFER.indication; Move to state AR
MultiplexEntrySendRelease	Do nothing

Current state=Awaiting Response (AR)	
Message, event, or primitive	Actions
TRANSFER.response	Send MultiplexEntrySendAck; Move to state IDLE
REJECT.request	Send MultiplexEntrySendReject; Move to state IDLE
MultiplexEntrySend	Set in_SQ=sequenceNumber; Inform REJECT.indication; Inform TRANSFER.indication
MultiplexEntrySendRelease	Inform REJECT.indication(PROTOCOL); Move to state IDLE

B.7   Request Multiplex Entry Signaling Entities (RMESEs)

B.7.1   Outgoing RMESE

Allowed states of the outgoing RMESE

State	Comment
IDLE	The state before and after the procedure has executed
Awaiting Response (AR)	The state immediately after sending a request

Outgoing RMESE internal state variables

Variable	Description	Comment
out_ENUM	Indicates the outgoing RMESE instance: (1...15)	Identifies the outgoing RMESE and related multiplexTableEntryNumber

Outgoing RMESE timers

Identifier	Description	Comment
T107	Limits the time the SE will wait for a response	Should be slightly longer than the round-trip time.

Primitives from SE user to outgoing RMESE

Primitive	Purpose
SEND.request	Initiates request multiplex table entry

Primitives from outgoing RMESE to SE user

Primitive	Purpose (and parameter default values)
SEND.confirm	Informs SE user of the success of the procedure
REJECT.indication (SOURCE, CAUSE)	Informs SE user of rejection (USER, null)

Messages sent by outgoing RMESE

Message	Type	Default values
RequestMultiplexEntry{ multiplexTableEntryNumber }	Request	multiplexTableEntryNumber = out_ENUM
RequestMultiplexEntryRelease{ multiplexTableEntryNumber }	Indication	multiplexTableEntryNumber = out_ENUM

Definition of outgoing RMESE operation. The Outgoing RMESE is initialized with the value of out_ENUM corresponding to the multiplexTableEntryNumber that the outgoing RMESE refers to.

Current state=IDLE	
Message, event, or primitive	Actions
SEND.request	Send RequestMultiplexEntry Request; Set timer T107; Move to state AR
RequestMultiplexEntryAck	Do nothing
RequestMultiplexEntryReject	Do nothing

Current state=Awaiting Response (AR)	
Message, event, or primitive	Actions
RequestMultiplexEntryAck	Reset timer T107; Inform SEND.confirm; Move to state IDLE
RequestMultiplexEntryReject	Reset timer T107; Inform REJECT.indication (RequestMultiplexEntry.rejectionDescriptions.cause); Move to state IDLE
Timer Expiry	Send RequestMultiplexEntryRelease; Inform REJECT.indication(PROTOCOL); Move to state IDLE

B.7.2   Incoming RMESE

Allowed states of the incoming RMESE

State	Comment
IDLE	The state before and after the procedure has executed
Awaiting Response (AR)	The state immediately after receiving a request

Incoming RMESE internal state variables

Variable	Description	Comment
In_ENUM	Indicates the incoming RMESE instance: (1...15)	Identifies the incoming RMESE and related multiplexTableEntryNumber

Primitives from SE user to incoming RMESE

Primitive	Purpose
SEND.response	Initiates Ack: accepts request
REJECT.request(CAUSE)	Initiates Reject: rejects request

Primitives from incoming RMESE to SE user

Primitive	Purpose (and parameter default values)
SEND.indication	Informs SE user of RequestMultiplexEntry request
REJECT.indication(SOURCE)	Informs SE user of rejection (USER)

Messages sent by the incoming RMESE

Message	Type	Default value

RequestMultiplexEntryAck{
multiplexTableEntryNumber
 	
}
	Response	multiplexTableEntryNumber
		= in_ENUM

RequestMultiplexEntryReject{
multiplexTableEntryNumber
cause
}
	Response	multiplexTableEntryNumber
		= in_ENUM
		cause=CAUSE

Definition of incoming RMESE operation. The Incoming RMESE is initialized with the value of in_ENUM corresponding to the multiplexTableEntryNumber that the outgoing RMESE refers to.

Current state=IDLE	
Message, event, or primitive	Actions
RequestMultiplexEntry	Inform SEND.indication; Move to state AR
RequestMultiplexEntryRelease	Do nothing

Current state=Awaiting Response (AR)	
Message, event, or primitive	Actions
SEND.response	Send RequestMultiplexEntryAck; Move to state IDLE
REJECT.request	Send RequestMultiplexEntryReject; Move to state IDLE
RequestMultiplexEntry	Inform REJECT.indication Inform SEND.indication
RequestMultiplexEntryRelease	Inform REJECT.indication(PROTOCOL); Move to state IDLE

B.8   Mode Request Signaling Entities (MRSEs)

B.8.1   Outgoing MRSE

Allowed states of the outgoing MRSE

State	Comment
IDLE	The state before and after the procedure has executed
Awaiting Response (AR)	The state immediately after sending a request

Outgoing MRSE internal state variables

Variable	Description	Comment
out_SQ	Identifies the most recently sent request: (0...255)	Associates a response with a particular request

Outgoing MRSE timers

Identifier	Description	Comment
T109	Limits the time the SE will wait for a response	Should be slightly longer than the round-trip time.

Primitives from SE user to outgoing MRSE

Primitive	Purpose
TRANSFER.request(MODE-ELEMENT)	Initiates request mode

Primitives from outgoing MRSE to SE user

Primitive	Purpose (and parameter default values)
TRANSFER.confirm (MODE-PREF)	Informs SE user of the success of the procedure (RequestModeAck.response)
REJECT.indication (SOURCE, CAUSE)	Informs SE user of rejection (USER, null)

Messages sent by outgoing MRSE

Message	Type	Default values
RequestMode{ sequenceNumber requestedModes }	Request	sequenceNumber = out_SQ requestedModes = MODE-ELEMENT
RequestModeRelease	Indication	—

Definition of outgoing MRSE operation. The value of out_SQ is not initialized so may start at any value.

Current state=IDLE	
Message, event, or primitive	Actions
TRANSFER.request	Increment out_SQ; Send RequestMode Request; Set timer T109; Move to state AR
RequestModeAck	Do nothing
RequestModeReject	Do nothing

Current state=Awaiting Response (AR)	
Message, event, or primitive	Actions

RequestModeAck	If sequenceNumber=out_SQ{
	  Reset timer T109;
	  Inform TRANSFERconfirm;
	  Move to state IDLE
	}
	Else{
	  Do nothing
	}

RequestModeReject	If sequenceNumber=out_SQ{
	  Reset timer T109;
	  Inform REJECT.indication(USER,
	RequestModeReject.cause);
	  Move to state IDLE
	}
	Else{
	  Do nothing
	}

TRANSFER.request	Reset timer T109; Increment out_SQ; Send RequestMode; Set timer T109
Timer Expiry	Send RequestModeRelease; Inform REJECT.indication(PROTOCOL, null); Move to state IDLE

B.8.2 Incoming MRSE

Allowed states of the incoming MRSE

State	Comment
IDLE	The state before and after the procedure has executed
Awaiting Response (AR)	The state immediately after receiving a request

Incoming MRSE internal state variables

Variable	Description	Comment
in_SQ	Indicates the most recently received request: (0...255)	Used to identify which request a response is associated with

Primitives from SE user to incoming MRSE

Primitive	Purpose
TRANSFER.response(MODE-PREF)	Initiates Ack: accepts request
REJECT.request(CAUSE)	Initiates Reject: rejects request

Primitives from incoming MRSE to SE user

Primitive	Purpose (and default parameter values
TRANSFER.indication(MODE-ELEMENT)	Informs SE user of RequestMode request (RequestMode.requestedModes)
REJECT.indication(SOURCE)	Informs SE User of rejection (USER)

Messages sent by the incoming MRSE

Message	Type	Default field values
RequestModeAck{ SequenceNumber }	Response	sequenceNumber = in_SQ
RequestModeReject{ SequenceNumber cause }	Response	sequenceNumber=in_SQ cause = CAUSE

Definition of incoming MRSE Operation

Current state=IDLE	
Message, event, or primitive	Actions
RequestMode	Set in_SQ=sequenceNumber; Inform TRANSFER.indication; Move to state AR
RequestModeRelease	Do nothing

Current state=Awaiting Response (AR)	
Message, event, or primitive	Actions
TRANSFER.response	Send RequestModeAck; Move to state IDLE
REJECT.request	Send RequestModeReject; Move to state IDLE
RequestMode	Set in_SQ=sequenceNumber Inform REJECT.indication; Inform TRANSFER.indication
RequestModeRelease	Inform REJECT.indication(PROTOCOL); Move to state IDLE

B.9   Round Trip Delay Signaling Entity (RTDSE)

Allowed states of the RTDSE

State	Comment
IDLE	The state before and after the procedure has executed
Awaiting Response (AR)	The state immediately after sending a request

RTDSE internal state variables

Variable	Description	Comment
out_SQ	Identifies the most recently sent request: (0...255)	Associates a response with a particular request

RTDSE timers

Identifier	Description	Comment
T105	Limits the time the SE will wait for a response	Should be two to three times the expected round-trip time.

Primitives from SE user to RTDSE

Primitive	Purpose
TRANSFER.request	Initiates round trip delay

Primitives from RTDSE to SE user

Primitive	Purpose (and parameter default values)
TRANSFER.confirm(DELAY)	Informs SE user of the success of the procedure (initial value of T105 minus value of T105)
EXPIRY.indication	Informs SE user of the failure of the procedure

Messages between RTDSE and peer RTDSE

Message	Type	Default values
RoundTripDelayRequest{ sequenceNumber }	Request	sequenceNumber=out_SQ
RoundTripDelayResponse{ sequenceNumber }	Response	sequenceNumber = RoundTripDelayRequest.sequenceNumber

Internal computations in RTSE

The value of DELAY is computed by comparing the value of the timer when a response is received to T105. DELAY=T105 − (timer value).

Definition of RTSE operation. out_SQ is not initialized so may start at any value.

Current state=IDLE	
Message, event, or primitive	Actions
TRANSFER.request	Increment out_SQ; Send RoundtripDelayRequest; Set timer T105; Move to state AR
RoundTripDelayResponse	Do nothing
RoundTripDelayRequest	Send RoundTripDelayResponse

Current state=Awaiting Response (AR)	
Message, event, or primitive	Actions

RoundTripDelayResponse	If sequenceNumber=out_SQ{
	  Inform TRANSFERconfirm;
	  Reset timer T105;
	  Move to state IDLE
	  }
	Else{
	  Do nothing
	  }

TRANSFER.request	Reset timer T105; Increment out_SQ; Send RoundTripDelayRequest; Set timer T105
Timer Expiry	Inform EXPIRY.indication Move to state IDLE
RoundtripDelayRequest	Send RoundtripDelayResponse

B.10   Maintenance Loop Signaling Entities (MLSEs)

B.10.1   Outgoing MLSE

Allowed states of the outgoing MLSE

State	Comment
Not Looped (NL)	The state when the Maintenance Loop is off
Awaiting Response (AR)	The state immediately after a request is sent
Looped (LPD)	The state when the Maintenance Loop is active

Outgoing MLSE internal state variables

Variable	Description	Comment
out_MLN	Maintenance Loop number: (0...255)	Identifies the outgoing MLSE

Outgoing MLSE timers

Identifier	Description	Comment
T102	The maximum time the SE will wait for a response	Should be slightly longer than the round-trip time

Primitives from SE user to outgoing MLSE

Primitive	Purpose
LOOP.request(LOOP_TYPE)	Initiates establishment of Maintenance Loop
RELEASE.request	Initiates command to switch loop off

Primitives from outgoing MLSE to SE user

Primitive	Purpose (and parameter default values)
LOOP.confirm	Informs SE user of successful establishment of loop
RELEASE.confirm	Informs SE user that loop has been switched off
RELEASE.indication (SOURCE, CAUSE)	Informs SE user of rejection (USER, MaintenanceLoopReject.cause)

Error primitives from outgoing MLSE to SE user

Error primitive	Error condition
ERROR.indication(A)	Inappropriate MaintenanceLoopAck message received
ERROR.indication(B)	T102 timer expiry: no response from remote incoming MLSE

Messages sent by outgoing MLSE

Message	Type	Default values
MaintenanceLoopRequest{ type }	Request	type = {LOOP_TYPE, out_MLN}
MaintenanceLoopOffCommand	Command	

Definition of outgoing MLSE operation.   out_MLN is initialized to the number of the logical channel it is associated with.

Current state=Not Looped (NL)	
Message, event, or primitive	Actions
LOOP.request	Send MaintenanceLoopRequest; Set timer T102; Move to state AR
MaintenanceLoopAck	Do nothing
MaintenanceLoopReject	Do nothing

Current state=Awaiting Response (AR)	
Message, event, or primitive	Actions
MaintenanceLoopAck	Reset timer T102; Inform LOOP.confirm; Move to state LPD
MaintenanceLoopReject	Reset timer T102; Inform RELEASE.indication; Move to state NL
RELEASE.request	Reset timer T102; Send MaintenanceLoopOffCommand; Move to state NL
Timer Expiry	Inform ERROR.indication(B); Send MaintenanceLoopOffCommand; Inform RELEASE.indication(MLSE, null); Move to state NL

Current state=Looped (LPD)	
Message, event, or primitive	Actions
RELEASE.request	Send MaintenanceLoopOffCommand; Move to state NL
MaintenanceLoopAck	Do nothing
MaintenanceLoopReject	Inform ERROR.indication(A); Inform RELEASE.indication(MLSE, null); Move to state NL

B.10.2   Incoming MLSE

Allowed states of the incoming MLSE

State	Comment
Not Looped (NL)	The state when the Maintenance Loop is off
Awaiting Response (AR)	The state immediately after receiving a MaintenanceLoopRequest
Looped (LPD)	The state when the Maintenance Loop is active

Incoming MLSE internal state variables

Variable	Description	Comment
in_MLN	Maintenance Loop number: (0...255)	Identifies the incoming MLSE
In_TYPE	Stores the value of LOOP_TYPE	Is set to MaintenanceLoopRequest.type in the incoming MaintenanceLoopRequest message

Primitives from SE user to incoming MLSE

Primitive	Purpose
LOOP.response	Initiate Ack: request successful
RELEASE.request(CAUSE)	Initiate Reject: request unsuccessful

Primitives from incoming MLSE to SE user

Primitive	Purpose (and parameter default values)
LOOP.indication (LOOP_TYPE)	Inform SE user of request (MaintenanceLoopRequest.type)
RELEASE.indication (SOURCE)	Informs SE user that loop is being switched off (USER)

Messages sent by incoming MLSE

Message	Type	Field values
MaintenanceLoopAck{ type }	Response	type = {in_TYPE, in_MLN}
MaintenanceLoopReject{ type cause }	Response	type = {in_TYPE, in_MLN} cause = CAUSE

Definition of incoming MLSE operation.   in_MLN is initialized to the number of the logical channel it is associated with.

Current state=Not Looped (NL)	
Message, event, or primitive	Actions
MaintenanceLoopRequest	Inform LOOP.indication; Move to state AR
MaintenanceLoopOffCommand	Do nothing

Current state=Awaiting Response (AR)	
Message, event, or primitive	Actions
LOOP.response	Send MaintenanceLoopAck; Move to state LPD
RELEASE.request	Send MaintenanceLoopReject; Move to state NL
MaintenanceLoopOffCommand	Inform RELEASE.indication; Move to state NL
MaintenanceLoopRequest	Inform RELEASE.indication; Inform LOOP.indication; Move to state AR

Current state=Looped (LPD)	
Message, event, or primitive	Actions
MaintenanceLoopOffCommand	Inform RELEASE.indication; Move to state NL
MaintenanceLoopRequest	Inform RELEASE.indication; Inform LOOP.indication; Move to state AR

Acronyms and Abbreviations

16CIF	A picture format with double the linear dimensions of 4CIF
2G	Second generation mobile network
3G	Third generation mobile network
3G-324M	A variant of H.324M specified by 3GPP
3GPP	Third generation partnership project for 3G UMTS
3GPP2	Sister body to 3PP for 3G systems not evolved from GSM
4CIF	A picture format with double the linear dimensions of CIF
ASIC	Application Specific Integrated Circuit
ADSL	Assymetric Digital Subscriber Loop
AL	Adaptation Layer
AMR	Adaptive Multirate (audio codec)
ARIB	Association of Radio Industries and Businesses
ASCII	American Standard Code for Information Interchange
ASN.1	Abstract Syntax Notation 1
ATIS	Alliance for Telecommunications Industry Solutions
ATM	Asynchronous Transfer Mode
AVP	Audio Video Profile
BC	Bearer Capability
BER	Basic Encoding Rules
BSC	Base Station Controller
BSS	BSC Subsystem
BTS	Base Tranceiver Station
CCSA	China Communications Standards Association
CCSRL	Segmentation and Reassembly Layer
CELP	Code Excited Linear Prediction
CESE	Capability Exchange Signaling Entity
CIF	Common Intermediate Format (352×288 pixels)
CLCSE	Close Logical Channel Signaling Entity
CRC	Cyclic Redundancy Check
DCT	Discrete Cosine Transform

DECT	Digital Enhanced Cordless Telecommunications
DNA	Dilithium Networks Analyzer
DSP	Digital Signal Processor DTRX Declined Retransmission
DTX	Discontinuous Transmission
EDGE	Enhanced Data Rates for GSM Evolution
EI	Error Indication
ETSI	European Telecommunications Standards Institute
FCS	Frame Check Sequence
FNUR	Fixed Network User Rate
FOMA	Freedom of Mobile Multimedia Access
FSM	Finite State Machine
G.711	ITU-T PCM speech standard at 64 kbit/s
G.723.1	ITU-T speech standard at 6.3 and 5.3 kbit/s
GGSN	Gateway GPRS Support Node
GMSC	Gateway MSC
GOB	Group of Blocks
GPP	General Purpose Processor
GPRS	General Packet Radio Service
GSM	Global System for Mobile Communication
GSM-AMR	GSM Adaptive Multirate speech codec
GSM-EFR	GSM Enhanced Full-rate speech codec
GSM-FR	GSM Full-rate speech codec
GSM-HR	GSM Half-rate speech codec
GSTN	General Switched Telephone Network
HEC	Header Error Control
HLR	Home Location Register
http	Hypertext Transfer Protocol
IE	Information Element
IETF	Internet Engineering Task Force
IMTC	International Multimedia Telecommunications Consortium
IN	Intelligent Network
IP	Internet Protocol
IPv4	Internet Protocol version 4
IPv6	Internet Protocol version 6
ISDN	Integrated Services Digital Network
ISO	International Standards Organization
ITC	Information Transfer Capability
ITU	International Telecommunications Union

ITU-R	Radiocommunication Sector of the ITU
ITU-T	ITU Telecommunications standardization sector
IVLC	Inverse VLC
LAN	Local Area Network
LCN	Logical Channel Number
LCSE	Logical Channel Signaling Entity
LLC	Low Layer Compatibility
LS	Last Segment
LSB	Least Significant Bit
MB	Macroblock
MC	Multiplex Code
MCU	Multipoint Control Unit
MGC	Multimedia Gateway Controller
MGW	Multimedia Gateway
MIME	Multipurpose Internet Mail Extension
ML	Mobile Level
MMS	Multimedia Messaging Service
MPEG	Moving Picture Experts Group
MPL	Multiplex Payload Length
MSB	Most Significant Bit
MSC	Mobile Switching Center
MSDSE	Master Slave Determination Signaling Entity
MTE	Multiplex Table Entry
MTSE	Multiplex Table Signaling Entity
MUX	Multiplexer
Node B	3G equivalent of a GSM BTS
NSRP	Numbered Simple Retransmission Protocol
PCM	Pulse Code Modulation
PDA	Personal Digital Assistant
PDU	Protocol Data Unit
PER	Packed Encoding Rules
PM	Packet Marker
PPP	Point to Point Protocol
PSC	Picture Start Code
PT	Payload Type
QCIF	Quarter CIF
QoS	Quality of Service
RAN	Radio Access Network

RC	Repeat Count
RFC	Request for Comment
RGB	Red Green Blue
RMESE	Request Multiplex Entry Signaling Entity
RNC	Radio Network Controller
RTCP	Real Time Control Protocol
RTDSE	Round Trip Delay Signaling Entity
RTP	Real Time Protocol
RVLC	Reversible Variable Length Coding
SDL	Specification and Description Language
SDP	Session Description Protocol
SDU	Service Data Unit
SE	Signaling Entity
SGSN	Serving GPRS Support Node
SID	Silence Insertion Descriptor
SigComp	Signaling Compression
SIP	Session Initiation Protocol
SMG	Special Mobile Group
SMS	Short Message Service
SQCIF	Sub-QCIF 128×96 pixel picture format
SREJ	Selective Reject
SRP	Simple Retransmission Protocol
SS7	Signaling System number 7
TCP	Transmission Control Protocol
TIA	Telecommunications Industry Association
TRAU	Transcoding and Rate Adaptation Unit
TTA	Telecommunication Technology Association
TTC	Telecommunications Technology Committee
UA	User Agent
UAC	User Agent Client
UAS	User Agent Server
UDI	Unrestricted Digital Information
UDP	User Datagram Protocol
UE	User Equipment
UI	User Interface
UMTS	Universal Mobile Telephony Service
URL	Uniform Resource Locator

VAD	Voice Activity Detection
VLC	Variable Length Coding
VLR	Visitor Location Register
VoIP	Voice over IP
VOP	Video Object Plane
WAIUR	Wanted Air Interface User Rate

Bibliography

3rd Generation Partnership Project (3GPP), TS 24.008: Mobile radio interface layer 3 specification; Core Network Protocols—Stage 3; www.3gpp.org.

3rd Generation Partnership Project (3GPP), TS 24.228: Signalling flows for the IP multimedia call control based on SIP and SDP; www.3gpp.org.

3rd Generation Partnership Project (3GPP), TS 24.229: IP Multimedia Call Control Protocol based on SIP and SDP; www.3gpp.org.

3rd Generation Partnership Project (3GPP), TS 25.301: Radio Interface Protocol Architecture; www.3gpp.org.

3rd Generation Partnership Project (3GPP), TS 26.235: Packet Switched Conversational Multimedia Applications; Default Codecs (Release 5); www.3gpp.org.

3rd Generation Partnership Project (3GPP), TS 26.911: Terminal Implementor's Guide; www.3gpp.org.

3rd Generation Partnership Project (3GPP), TS 26.110: Codec(s) for Circuit Switched Multimedia Telephony Service: General Description; www.3gpp.org.

3rd Generation Partnership Project (3GPP), TS 26.111: Codec(s) for Circuit Switched Multimedia Telephony Service, Modifications to H.324; www.3gpp.org.

3rd Generation Partnership Project (3GPP), TS 26.236: Packet switched conversational multimedia applications; Transport protocols (Release 5); www.3gpp.org.

3rd Generation Partnership Project (3GPP), TS 27.001: General on Terminal Adaptation Functions (TAF) for Mobile Stations (MS) (Release 99); www.3gpp.org3rd Generation Partnership Project 2 (3GPP2), S.R0022: Video Conferencing Services—Stage 1; www.3gpp.org.

Future mobile networks. *B.T. Technol. J.*, Vol. 19, no. 1, January 2001.

Gonzalo Camarillo: SIP Demystified., McGraw-Hill, 2002.

GSM 06.73: ANSI-C code for the GSM adaptive multi rate speech codec. (Also issued as 3GPP TS 26.073; www.3gpp.org)

GSM 06.90: Adaptive multi rate speech transcoding. (Also issued as 3GPP TS 26.090; www.3gpp.org)

GSM 06.91: Substitution and muting of lost frames for adaptive multi rate speech traffic channels. (Also issued as 3GPP TS 26.091; www.3gpp.org)

GSM 06.92: Comfort noise aspects for adaptive multi rate speech traffic channels.

GSM 06.94: Voice activity detector for adaptive multi rate speech traffic channels.

Harri Holma and Antti Toskala (eds.): WCDMA for UMTS, 2nd ed. John Wiley & Sons, 2002.

Iain E.G. Richardson: H.264 and MPEG-4 Video Compression: Video Coding for Next Generation Multimedia. Halsted Press, 2003.

ISO/IEC: Information Technology—Coding of Audio-Visual Objects. Part 2: *Visual*, ISO/IEC 14496-2, 1999.

ITU-T: Dual rate speech coder for multimedia communication transmitting at 5.3 and 6.3 kbit/s. ITU-T Recommendation G.723.1, 1996; www.itu.org.

ITU-T: Multiplexing protocol for low bit-rate multimedia communication. ITU-T Recommendation H.223, 2001; www.itu.org.

ITU-T: Control protocol for multimedia communication. ITU-T Recommendation H.245, 2000; www.itu.org.

ITU-T: Video coding for low bit-rate communication. ITU-T Recommendation H.263, 1998; www.itu.org.

ITU-T: Narrow-band visual telephone systems and terminal equipment. ITU-T Recommendation H.320, 1997; www.itu.org.

ITU-T: Packet-based multimedia communication systems. ITU-T Recommendation H.323, 1998; www.itu.org.

ITU-T: Terminal for low bitrate multimedia communication. ITU-T Recommendation H.324, 1998; www.itu.org.

ITU-T: Digital subscriber signalling system no. 1 (DSS1)—ISDN usernetwork interface layer 3 specification for basic call control. ITU-T Recommendation Q.931.1998; www.itu.org.

ITU-T: Information Technology—Abstract Syntax Notation One (ASN.1)—Specification of basic notation. ITU-T Recommendation X.680, 1994; www.itu.org.

ITU-T: Information Technology—ASN.1 Encoding Rules—Specification of Packed Encoding Rules (PER). ITU-T Recommendation X.691, 1996; www.itu.org.

Mohammed Ghanbari: Standard Codecs: Image Compression to Advanced Video Coding. IEE Publishing, June 2003.

RFC 2327: SDP: Session Description Protocol. April 1998; www.ietf.org.

RFC 3113: 3GPP-IETF Standardization Collaboration. June 2001; www.ietf.org.

RFC 3261: SIP: Session Initiation Protocol. June 2002; www.ietf.org.

RFC 3262: Reliability of provisional responses in Session Initiation Protocol (SIP). June 2002; www.ietf.org.

RFC 3264: An Offer/Answer Model with SDP. June 2002; www.ietf.org.

RFC 3311: The Session Initiation Protocol (SIP) UPDATE method. September 2002; www.ietf.org.

RFC 3320: Signaling Compression (SigComp). January 2003; www.ietf.org.

RFC 3486: Compressing the Session Initiation Protocol (SIP). February 2003; www.ietf.org.

RFC 3551: RTP Profile for Audio and Video Conferences with Minimal Control. July 2003; www.ietf.org.

Index

ABOUT THE AUTHOR

DAVID J. MYERS is a co-founder of Dilithium Networks and was its first Vice President of Engineering. A multiyear veteran of British Telecom's R&D division, he served as Broadband Delivery Manager for BTopenworld, the largest broadband provider in the United Kingdom. He has also lectured on telecommunications engineering at Sydney University, and acted as a technical consultant to the European Commission.